Clinicians' Guide to Aspirin

Clinicians' Guide to Aspirin

Antony Bayer MB
University Department of Geriatric Medicine,
University of Wales College of Medicine,
Llandough Hospital,
Penarth,
Wales

CHAPMAN & HALL MEDICAL

Published by Chapman & Hall, an imprint of Lippincott-Raven Publishers, 2–6 Boundary Row, London SE1 8HN, UK

Lippincott-Raven Publishers, 227 East Washington Square, Philadelphia PA 19106-3720, USA

First edition 1998

Typeset in 11/13pt Adobe Garamond by Photoprint, Torquay, Devon
Printed in Great Britain at the Alden Press, Oxford

ISBN 0 412 81900 7

A catalogue record for this book is available from the British Library

∞ Printed on acid-free text paper, manufactured in accordance with ANSI/NISO Z39.48-1992 (Permanence of Paper).

Contents

Colour plates appear between pages 14 and 15

Preface

Aspirin is unique. There are few drugs as long established, as well known or as widely used. Nor are there many agents as efficacious in reducing morbidity and mortality in so many different clinical settings. It has led to the development of a whole family of other non-steroidal anti-inflammatory drugs, revolutionized secondary prevention of vascular disease, provided insights into the mechanisms underlying inflammation, thrombosis and certain cancers, played a major role in the development of the new statistical science of meta-analysis and, 100 years after its introduction into clinical practice, is still the subject of many hundreds of scientific papers each year.

The centenary of aspirin provides an opportunity to consider its past, present and likely future role in medical practice. This book aims to review the accumulated evidence of the last century and to discuss its relevance to everyday clinical situations. It provides a concise reference to all those seeking information on its development, its pharmacological background and current therapeutic indications. All drugs have potential adverse effects and these are considered in the final chapters.

Antony Bayer, Cardiff

Acknowledgements

I am pleased to acknowledge the help given by Bayer AG, Leverkusen, in providing the charming illustrations of historical Aspirin advertisements and for their permission to reproduce these in this book.

I would also like to acknowledge the influence and infectious enthusiasm of Professor Peter Elwood, who ignited my academic and research interest in aspirin, first kindled by my involvement in recruiting patients into the ISIS-2 and IST studies in acute myocardial infarction and acute stroke. Despite my surname, I have no other association with aspirin though my brother, Nicholas Bayer, had been nicknamed 'Aspro' by the medical and nursing staff when he was born!

I am indebted to my secretary, Heather Copeland, for her calm efficiency and consistent support and am grateful for the help of the Cochrane Library and Medical Illustration Department of Llandough Hospital and to the Welsh Poisons Information Service. On a personal note, I thank my wife Eileen and Rachel, Nick, Sam, Ben and Oliver for their patience and understanding during the writing of this book.

Historical background

Introduction

Acetylsalicylic acid was first synthesized and introduced into clinical practice at the end of the 19th century by the German chemical company Bayer. With an appreciation of the marketing value of a short and memorable name, it was rechristened aspirin. This is generally accepted to have been derived from the product's German name, *acetylspirsaure* (so called because salicylate was obtained from plants of the genus *Spiraea*) and the suffix *-in*. A more colourful explanation is that it was named after Saint Aspirinius, an early Neapolitan bishop who was the patron saint against headaches [1].

Aspirin soon became a household name around the world. As a result of Germany's defeat in the Great War, Bayer lost exclusive rights to its use in Britain and France and in America a court case ruled that the word had entered the public domain. With numerous pharmaceutical companies moving into production and marketing, the numbers of people regularly taking aspirin as an analgesic, antirheumatic and antipyretic rose into the tens of millions. In the second half of the century, the clinical potential of aspirin's antithrombotic properties was recognized and it acquired a central role in the prevention and treatment of ischaemic heart disease and cerebrovascular disease.

Aspirin accompanied the first expedition to conquer Everest and the Apollo 11 astronauts on the first manned mission to the moon. During prohibition in America it briefly acquired an unjustified reputation as a potential stimulant drug when taken with Coca Cola. It may also have played a central role in shaping world history, by helping Rasputin to establish his influence in the Russian Court through the success of his advice not to give aspirin to treat the joint pains of the haemophiliac son of the Tsar [2]. It may even have helped to prolong the life of the stroke-prone Winston Churchill, who regularly took low-dose aspirin in combination with amphetamines while Prime Minister in the 1940s [3].

One hundred years after its introduction into clinical practice, scientific interest in aspirin continues to grow (Figure 1.1). A critical biographic

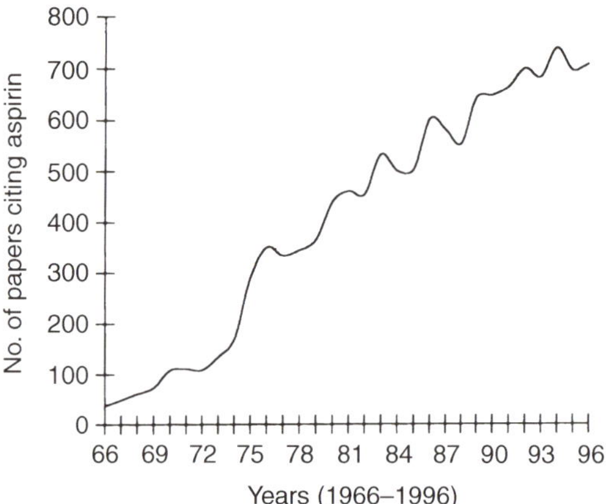

Fig. 1.1
Citations of aspirin in *Index Medicus* each year, from 1966 to 1996

review of the literature on the salicylates published in 1948 listed about 4000 references on aspirin [4]. Fifty years later, a Medline search will identify a similar number published in the last five years alone.

Early history

IDENTIFICATION OF SALICYLATE-CONTAINING PLANTS

The therapeutic uses of the salicylates have a much longer history than just the last 100 years [5–9]. Salicylate-containing plants, such as the willow (*Salix*) and myrtle and poplar (*Spiraea*), were used by the ancient Egyptians, the Greeks and the Romans to relieve pain and reduce fever and were traditional folk remedies for a range of ailments right up to the last century (Figure 1.2).

Meadowsweet (*Spiraea ulmaria*) was a common constituent of English herb gardens. All contain salicin, the natural glycoside from which salicylate may be extracted. Methyl salicylate may also be found in the Wintergreen family (*Gaultheria*) and small amounts are present in hawthorn, beech and birch trees, liquorice and olives and cherry, plum, grape, strawberry, raspberry, orange and apple. Even the flavour characteristics of tea are influenced by its methyl salicylate content [9]. Numerous bacteria produce salicylic acid, which is involved with iron transport and control of growth. There are only two identified animal sources, the oily secretion from the scent glands of beavers, known as castor [4], and a defence secretion in the ponerine ant [9].

Fig. 1.2
Seventeenth century woodcuts of the shrub meadowsweet (*Spiraea ulmaria*) and of the white willow (*Salix alba*), both natural sources of salicylate (from reference [7] with permission).

CLINICAL USE IN ANCIENT CIVILIZATIONS

The earliest reference to the analgesic properties of salicin appears in a papyrus scroll discovered by Georg Ebers at Luxor in 1862. This is thought to date from about 1550 BC, but the 800 listed remedies probably originated from a much earlier dynasty. Amongst the suggested cures is a decoction of dried myrtle leaves prepared with beer and applied to the abdomen and back to dispel rheumatic pains from the womb [1].

Across the Mediterranean and 1000 years later, the father of modern medicine Hippocrates recommended the use of the bark and leaves of the willow tree to combat the pain of childbirth and to treat fever. He also promoted the juice of the poplar tree as treatment for eye diseases [7]. The anti-inflammatory and analgesic properties of extracts of willow and poplar are also mentioned in the works of the Roman encyclopaediatrist Celsus (boiled in vinegar to treat prolapsed uterus), the Roman naturalist and scholar Pliny the Elder (for sciatic pain and gout and as a paste for treating corns and callosities) and the Greek-educated physician Galen (to treat ulcerating wounds) [5].

Dioscorides, the first century Greek author of *De Materia Medica*, travelled widely as a surgeon in the Roman army. His authoritative five-volume pharmacopoeia contained all the pharmacological knowledge available in the empire and remained influential until the Renaissance. It recommends a decoction of willow leaves and bark as an excellent remedy for gout and earache and, when beaten small and taken with a little pepper and wine, as an effective cure for '*iliaca passio*' (presumably menstrual pain or colic) [8].

MEDIEVAL TIMES

By the Middle Ages, poultices and decoctions of salicylate-containing plants seem to have become established folk remedies for all forms of rheumatism, menstrual pains, migraine, wounds and ulcers and even dysentery. Rufinus, the early medieval physician, mentions the therapeutic benefits of willow trees and the *New Herbal* written by Dodoens in 1595 recommends 'The leaves and rinds of Withy boiled in wine do appease the paine of the sinuwes and do restore againe their strength, if they be nourished with the fomentation of natural heat thereof' [8]. A Swiss herbal compendium of 1486 recognized that application of willow flowers made into a plaster with rose oil would cause tissues that were hot and wet to become cold and dry, thus tacitly acknowledging its anti-inflammatory properties [9].

Some of the benefits claimed were clearly excessive. One 17th century French physician recommended preparations of willow leaves for the treatment of 'spitting of blood and all other fluxes of blood whatsoever in man or woman' [5].

An appreciation of the therapeutic potential of willow and *Spiraea* plants was not confined to Europe. They were certainly used by the South African Hottentots (the subject of a letter to *The Lancet* in 1876 [10]), the North American Indians and the ancient dynasties of Persia and China. The Persian physician Al-Samarqandi, writing in his *Medical Formulary* in the 13th century, recommended poultices and decoctions from the Egyptian willow (*Salix aegyptiaca*) and also mentioned that the weeping willow (*Salix babylonica*) was effective in treating migraine and other headaches [11]. In Burma, the bark of *Salix purpura* was used to treat rheumatism and in China preparations of willow and poplar were used for treatment of colds, haemorrhages, goitre and as a general antiseptic [9].

The Reverend Mr Edward Stone

An English clergyman, the Reverend Edward Stone, is credited with being the first to systematically investigate the properties of willow bark, in the mid-18th century [12]. He seems to have had wide interests, ranging from acting as a political agent for Whig party candidates in the general election of 1754 to writing a scientific paper in 1761 which identified where in the world would be the best place to see the impending passage of the planet Venus across the sun. This was taken up by the Admiralty in London, who dispatched Captain James Cook to Tahiti and subsequently led to the discovery of Australia [13].

Stone had led an academic life at Wadham College, Oxford, until he left in his early 40s to get married. His wife came from a local land-owning family and they eventually settled in the Oxfordshire town of Chipping Norton, acquiring a town house and about 20 acres of marshy land on the outskirts. As the family chaplain to a local nobleman, he no longer had extensive parish responsibilities and so instead had the opportunity to indulge himself in scientific enquiry.

By the mid-18th century, supplies of the quinine-containing cinchona (Peruvian or Jesuits') bark, widely used to treat malaria and other 'agues' (fever and rigors), were becoming scarce. The natural habitat of the cinchona was in the South American Andes and harvesting of the bark and export across the Atlantic were difficult and expensive. Early attempts to cultivate cinchona trees nearer to home were unsuccessful (although later large farms would be established in South-East Asia) and across Europe there was an urgent search for an effective substitute.

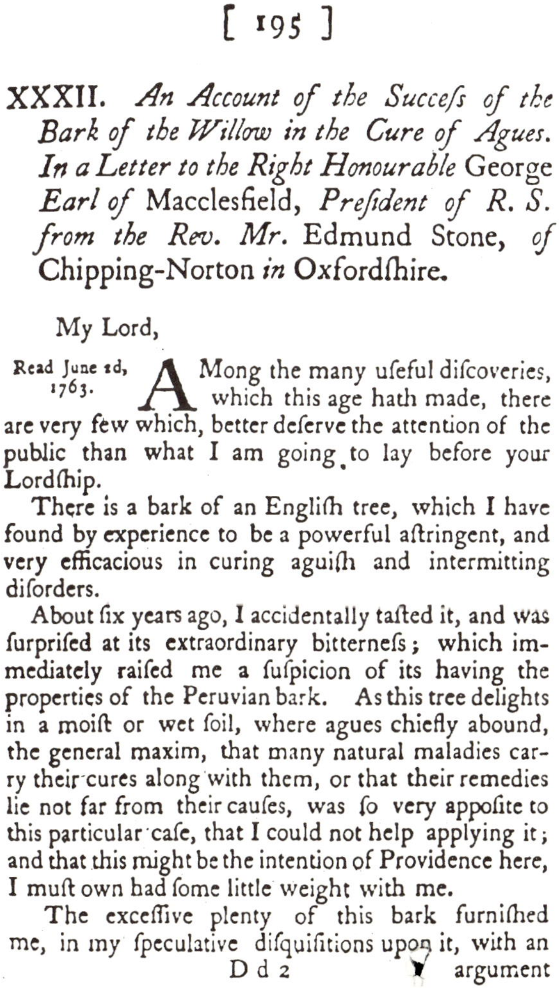

[195]

XXXII. *An Account of the Succefs of the Bark of the Willow in the Cure of Agues. In a Letter to the Right Honourable* George *Earl of* Macclesfield, *Prefident of R. S. from the Rev. Mr.* Edmund Stone, *of* Chipping-Norton *in* Oxfordfhire.

My Lord,

Read June 2d, 1763.

AMong the many ufeful difcoveries, which this age hath made, there are very few which, better deferve the attention of the public than what I am going to lay before your Lordfhip.

There is a bark of an Englifh tree, which I have found by experience to be a powerful aftringent, and very efficacious in curing aguifh and intermitting diforders.

About fix years ago, I accidentally tafted it, and was furprifed at its extraordinary bitternefs; which immediately raifed me a fufpicion of its having the properties of the Peruvian bark. As this tree delights in a moift or wet foil, where agues chiefly abound, the general maxim, that many natural maladies carry their cures along with them, or that their remedies lie not far from their caufes, was fo very appofite to this particular cafe, that I could not help applying it; and that this might be the intention of Providence here, I muft own had fome little weight with me.

The exceffive plenty of this bark furnifhed me, in my fpeculative difquifitions upon it, with an

D d 2 argument

Fig. 1.3
Front page of Edward Stone's report 'The Success of the Bark of the Willow in the Cure of Agues' published in the Philosophical Transactions of the Royal Society of London in June 1763 (from reference [12] with permission).

The Doctrine of Signatures suggested that plants growing in a particular area would cure the illnesses of its inhabitants, the local climate supposedly being responsible for both disease and cure. Stone noted that the willow 'delights in a moist or wet soil where agues chiefly abound' and recalled 'the general maxim that many natural maladies carry their cures along with them'. Impressed by the extraordinary bitterness of the bark of the white willow (*Salix alba*), reminiscent of the costly Peruvian bark, he determined to experiment with it. He dried some bark outside a baker's oven for three months, made it into a powder and over the next five years administered it 'successively and successfully' to about fifty persons suffering from agues and intermitting disorders. He claimed that it 'never failed in the cure, except in a few autumnal and quartan agues, with which the patients had been long and severely afflicted'.

In April 1763, Stone reported 'The Success of the Bark of the Willow in the Cure of Agues' in a letter to the Earl of Macclesfield, the President of the Royal Society of London (Figure 1.3). It was subsequently read out at a meeting of the Society and published in their *Philosophical Transactions* [12].

The 19th century

In accordance with current good research practice, Stone appears to have recognized the limited statistical power of his own studies and that a single clinical trial could not conclusively prove benefit. He concluded that his discovery needed 'a fair and full trial in all its variety and circumstances and situations'.

Unfortunately this suggestion does not seem to have attracted much interest, although numerous other case reports of the medicinal properties of willow continued to be published intermittently for the next century. It seems to have been fairly widely adopted as a treatment for fever and in 1798 an apothecary from Bath named William White was able to report that the substitution of willow bark for cinchona in the City Infirmary was saving the Charity at least £20 a year [9].

THOMAS MACLAGAN'S PAPER IN *THE LANCET*

Over 100 years after Stone's report, a Scottish physician, Dr Thomas MacLagan, published a paper in *The Lancet* of March 1876 reporting the efficacy of salicin in eight patients with rheumatic fever [14]. He claimed that the drug worked by dislodging the supposedly causative parasite which he thought was lodged in the heart and joints and responsible for the symptoms. Although he claimed 'no haphazard experiment' but one which 'had a fair foundation in reason and analogy', his work seems to have been once again inspired by a belief in the Doctrine of Signatures. *The Lancet* paper attracted considerable attention and MacLagan was able to move to London and build up a very successful practice. Demand for salicin was such that its price rose from two shillings to 12 shillings an ounce (comparable to that of quinine) and it was replaced by the cheaper synthetic salicylic acid which was only one shilling an ounce [15].

Salicylates were subjected to formal trial at several London teaching hospitals in the late 1870s and became the standard treatment for rheumatic fever. A report of the uricosuric properties of salicylate [16] encouraged its use in other conditions, including gout, neuralgia and even diabetes. Sodium and lithium salicylate were advocated in preference to salicylic acid as they were found to be more soluble and deemed to be better tolerated.

USE OF SALICYLATES IN EUROPE

About the same time as MacLagan's report, a biochemist, Marcellus von Nencki of Basle, had shown that salicin is converted in the body into salicylic acid. Also in Switzerland, Carl Buss then gave salicylic acid to typhoid patients as an internal antiseptic and noted that although it did not cure the infection, it was highly effective at lowering body temperature [17]. He also described occurrence of gastric irritancy and of tinnitus when high doses were used. In Berlin, Doctors Ludwig Riess and Franz Stricker independently reported the antipyretic and analgesic properties of sodium salicylate and salicylic acid, presenting in 1875 a series of over 400 patients mainly suffering from typhoid fever [18] and in the following year, evidence of its value in treating acute rheumatic fever [19]. It was suggested that it was less efficacious in chronic rheumatic conditions, such as rheumatoid and gonococcal arthritis. In France, however, Germain See reported great benefits in 12 cases of chronic rheumatism 'even after years of pain, stiffness and immobility' and was especially impressed by its prompt action in acute gout and its protective value in chronic gout such that 'the cure can be complete' [20].

The Paris correspondent of the *British Medical Journal* reported in 1877:

> The pharmaciens are also making a good harvest, and they offer the new remedy to the public in every variety of form as a panacea for the affections for which they are and are not indicated; but an unwitting public easily gulled, soon find to their cost that not only they have not been cured, but, terrified by some of the disagreeable symptoms produced by the medicine, give it up without allowing it a fair trial, and thus throw discredit on a remedy which, if properly and

intelligently employed may be looked upon almost as a specific treatment for gout and rheumatism [21].

RECOMMENDED DOSAGE

Certainly the doses being promoted in all these studies seem to have been very large. Stone's usual dose of 20–40 g of dried and powdered willow bark every four hours [12] would have resulted in a daily dose of between 5 g and 25 g of active salicin derivatives. MacLagan in his initial report [14] used 12 grains of salicin every three hours (a daily dose of 6 g) but recommended that doses 2–3 times greater could be given 'without experiencing the least inconvenience or discomfort'. See noted symptoms of 'deafness, tinnitus aurium and a peculiar cerebral disturbance' associated with his initial doses of 8–10 g/day of sodium salicylate and subsequently reduced the dose to 4 or 5 g/day [20]. Such high doses must have led to frequent adverse effects and it has been speculated that Ludwig van Beethoven's fatal renal papillary necrosis [22] may have been the consequence of salicin abuse.

ISOLATION OF SALICYLATE COMPOUNDS FROM PLANTS

Despite widespread use, it was well into the 19th century before chemical techniques were sufficiently advanced in continental Europe to allow the isolation and purification of the active principle of the willow bark. This achievement is variously awarded to Henri Leroux, a Parisian pharmacist, to Johann Buchner, a chemist working at the Pharmacological Institute of Munich, or to Fontana and Brugnatelli, working in Italy [23–25]. All isolated small amounts of the glycoside salicin from preparations of willow (*Salix*) in the late 1820s. A few years later, Raffaele Piria, an Italian chemist working in Paris, isolated salicylic acid by hydrolysis of the willow-derived glycoside.

In parallel with these studies, chemists working elsewhere in Europe were studying other salicylate-containing plants. In 1831, Johann Pagenstecher, a Swiss pharmacist, distilled salicylaldehyde from meadowsweet (*Spiraea ulmaria*) flowers. Soon afterwards, a Berlin chemist, Karl Jakob Lowig, obtained salicylic acid by oxidation of Pagenstecher's distilled aldehyde and called it 'spirsaure' (acid of *Spiraea*) because of its origin. Across the Atlantic in 1843, the American William Procter isolated methyl salicylate from oil of wintergreen (*Gaultheria procumbens*) which by hydrolysis yielded salicylic acid.

Thus by the 1850s there were three possible routes to isolate salicylate from plant material (Figure 1.4).

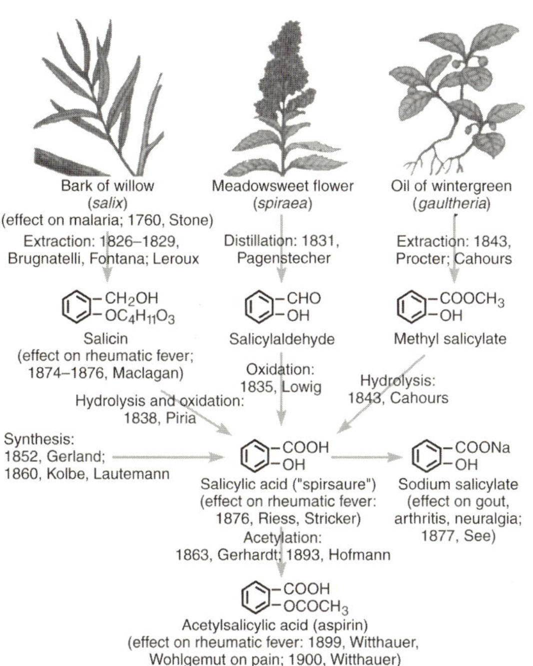

Fig. 1.4
The origins of aspirin: plant precursors, chemical inter-relations and pioneers (from reference [24] with permission).

FIRST SYNTHESIS

The first successful synthetic production of acetylsalicylic acid by the treatment of sodium salicylate with acetyl chloride is credited to an eminent French chemist, Charles Frederich von Gerhardt, in 1853. However, the significance of the discovery seems to have been largely unappreciated [26]. In the 1860s and 1870s, the German chemists Hermann Kolbe and Lautermann devised a practical method for the large-scale production of salicylic acid from phenol and carbon dioxide. In 1874 one of their students, Friedrich von Heyden, opened a factory in Dresden for its commercial production and to meet the demand from all over the world. By coincidence, Kolbe himself subsequently went to work for the Bayer Chemical Company [25]. Essentially it is Kolbe's original method which is still used in the commercial production of salicylic acid today, which is then acetylated with acetic anhydride to produce aspirin [9].

As well as its antirheumatic and antipyretic uses, salicylate was promoted as a surgical antiseptic (as an alternative to carbolic acid), as a topical keratolytic for treatment of warts, as a preservative for beer, milk and food and as a flavouring for soft drinks and sweets [7]. However, internal use of the purified salicylic acid and its sodium salt continued to be limited by the frequency of severe gastric intolerance. There was a need for a salicylate compound with a more favourable toxicity profile.

Felix Hoffmann and the Bayer Company

The Friedrich Bayer Company of Elberfeld in Germany produced aniline-based dyes from coal tar, as well as having a growing pharmaceutical division which had just successfully introduced heroin, misguidedly promoting it as a safe alternative to morphine. In 1895, a 29-year-old employee, Felix Hoffmann, was reassigned from the dye division to find and develop a better tolerated salicylate [1]. He was motivated in his search by the fact that his arthritic father could no longer tolerate salicylic acid because after taking it for many years it now caused him to vomit [8]. Hoffmann looked through the scientific literature and identified the work of Gerhardt from some 40 years before. On October 10th 1897, Hoffmann described in his laboratory notebook an improved synthetic pathway for producing acetylsalicylic acid, by acetylating the hydroxyl group on the benzene ring of salicylic acid. After first testing the compound on animals and himself, Hoffmann reputedly gave it to his father who is said to have benefited greatly.

Bayer's director of pharmacological research, Heinrich Dreser, initially dismissed the new compound as potentially cardiotoxic, but following glowing reports from independent clinical trials, he seems to have promoted it enthusiastically [6]. Indeed, his published report of pharmacokinetic studies on himself [27] fails to give any credit to Hoffmann or his colleagues. Furthermore, following a series of experiments which included dipping the tails of goldfish into various drug solutions, Dreser wrongly suggested that aspirin was merely a prodrug for salicylic acid, with a much less damaging effect on the stomach [9].

On February 1st 1899, the Bayer Company patented their new name for the drug. After considering 'a-salicin' and 'eusparin', they eventually chose aspirin, so that it would not be confused with salicylic acid. The Patent Office refused to register a patent for the discovery of the drug as they did not consider the acetylation of salicylic acid to be sufficiently novel. Thus Hoffmann failed to receive any royalties from his five years work [1]. Instead, the company was able to patent the processes used in its large-scale manufacture and enjoyed a monopoly on production of the drug for the next 17 years.

Aspirin was launched commercially in June 1899, first in the form of a white powder, then in gelatin capsules and eventually as compressed tablets. To ensure its successful launch, the company circulated more than 30 000 doctors in what was probably the first mass mailing of product information [17]. Aspirin very rapidly displaced alternative salicylate preparations, as well as competitors such as phenacetin and antipyrine. Dreser's contract with the Bayer Company awarded him a bonus on any product he successfully introduced and so he retired early, a very rich man.

FIRST REPORTS OF CLINICAL USE OF ASPIRIN

The first clinician to publish a report of the clinical use of aspirin was Kurt Witthaeuer from the Deaconess Hospital in Halle, who administered a dose of 1 g, 4–6–hourly in 50 patients with rheumatic fever, gout and pleurisy [28]. Despite initial scepticism, he reported favourable experiences and, in language reminiscent of the end of the 20th century, hoped that the medication would not be introduced at such a high price that it would prohibit general use. Another German physician, Julius Wohlegemut, administered aspirin 1–3 g to 10 patients with rheumatic diseases in his Berlin clinic and confirmed the findings [29]. The first English language report of the use of aspirin was by a Texan physician, F.C. Floeckinger, who reported that 'in all cases of acute and chronic muscular rheumatism the remedy proved extremely serviceable' [30].

The new drug went into widespread use throughout Europe and was introduced into America in 1915. It became the standard treatment for rheumatism and as early as 1905, a review article in *The Lancet* recommended that aspirin should be employed in every case [31].

NEW FORMULATIONS OF ASPIRIN

With the outbreak of the First World War, the British government became concerned that supplies of aspirin from Germany would soon run out. Ignoring the still valid Bayer patents, they launched a competition with a £20 000 prize to anyone in Britain or the Commonwealth who could develop a simple method of manufacturing the drug. A pharmacist, George Nicholas, living in Melbourne, Australia, won the prize and started to produce 'Aspro' tablets. At the end of the war, the British Custodian of Enemy Property sequestered the name 'aspirin' and the Bayer Company lost its exclusive rights to use the name in Britain and the Commonwealth [5].

In America, a dozen chemical firms led by the Monsanto Chemical Works began to manufacture aspirin [9]. In November 1918, Bayer lost the registered rights to the name of aspirin on the grounds that the patent had been unlawfully registered. The US government seized the firm's assets as belonging to enemy aliens and auctioned off the rights in the company to a small Virginian patent medicine firm, Sterling Products (later to become Sterling Winthrop). Three years later, they lost control over the drug when it was ruled that 'aspirin' had been so overadvertised that it had become a household name and therefore had entered the public domain [7].

Thus in Britain, France and the United States aspirin has long been a generic term and is marketed as such by numerous different manufacturers. Elsewhere in the world, Bayer AG has retained trademark rights and Bayer Aspirin is the only formulation available in over 70 countries [6].

With tens of millions of people taking aspirin worldwide for its analgesic, anti-inflammatory and antipyretic properties, reports of its adverse effects, especially gastric irritation, began to cause concern. Attempts to buffer the effects by combination with antacids were made and there was especial interest in the much more soluble calcium aspirin, which was recommended in Germany as early as 1913. It was not until the late 1930s, however, that the newly developed gastroscope was used to directly study the effects on the stomach lining of ordinary aspirin and calcium aspirin and the superiority of soluble preparations was fully appreciated [32,33]. Problems with stability limited the practical value of the newer preparations until the Hull-based company Reckitt and Colman developed a method of mass producing a soluble and stable version of calcium aspirin. It was successfully launched in 1948 with the slogan 'Take an aspirin – I mean a Disprin' [5].

Antithrombotic effects and prevention of vascular events

The antithrombotic effects of aspirin were recognized in the early 1940s by Karl Link who had earlier discovered the anticoagulants warfarin and dicoumarol. His finding that dicoumarol spontaneously broke down into salicylic acid prompted him to investigate the hypoprothrombinaemic action of aspirin [34]. Some years later, Armand Quick observed that aspirin ingestion prolonged the bleeding time and suggested that it depressed the plasma 'antibleeding factor' [35].

The clinical potential of this apparent antithrombotic effect was first investigated by Paul Gibson, who wrote an article in *The Lancet* in 1949 proposing the use of an oral solution of aspirin together with ascorbic acid in myocardial infarction and angina [36]. Meanwhile, a Californian ear, nose and throat specialist, Lawrence L. Craven, noted the bleeding associated with excessive use of an aspirin-containing analgesic gum prior to tonsillectomy and made the extraordinary but prescient observation that aspirin might have a preventive role in coronary thrombosis [37]. In two subsequent papers published in the somewhat obscure *Mississippi Valley Medical Journal*, he reported 100% protection against first and recurrent myocardial infarction and stroke in a series of over 8000 subjects taking aspirin prophylaxis over a period of eight years [38,39]. Unfortunately, despite following his own advice to take aspirin as a primary preventive measure, Craven himself seems to have been the exception to his rule and died of a myocardial infarct in 1957 [5].

It was not until the 1960s that the effect of aspirin on the haemostatic properties of platelets was described, providing a rational explanation for the cardioprotective effects [40]. The platelet effects were further reported in America by Harvey Weiss [41] and in Britain by John O'Brien [38] and prompted Peter Elwood and colleagues in the Medical Research Council Epidemiology Unit in Cardiff to embark on a randomized controlled trial of aspirin in the secondary prevention of mortality in 1200 patients who had recently been discharged from hospital following myocardial infarction [43]. This showed a non-significant but potentially very important 25% reduction in deaths after one year and encouraged numerous further large trials. One of the most influential was the ISIS-2 Study [44] which effectively transformed medical attitudes overnight. A survey in 1987, before the study was reported, showed that only 8% of British doctors used aspirin in the routine medical management of myocardial infarction. By 1996, this figure had risen to 85% [45].

The director of the MRC Unit and the second author of this first prophylactic aspirin trial was Archie Cochrane. His name is immortalized in the Cochrane Collaboration, which has become a worldwide movement responsible for statistical overviews of randomized controlled clinical trials. It was the published overviews of trials of aspirin [46] which established beyond doubt the major benefits of aspirin in secondary prevention not only of ischaemic heart disease but also of stroke. The most recent overview is now based on over 180 randomized controlled trials of antiplatelet therapy, involving 30 000 vascular events in 340 000 subjects.

Mechanism of action

Meanwhile, the mechanism of action of aspirin was unravelled by Sir John Vane and his colleagues working in the pharmacology laboratories of the Royal College of Surgeons in London. In a series of consecutive articles published in *Nature* in 1971, Vane demonstrated that aspirin and related drugs inhibited the synthesis of prostaglandins from arachidonic acid in guinea-pig lung homogenates [47], whilst his colleagues, Bryan Smith and Jim Willis, reported selective inhibition of prostaglandin synthetase in platelets [48] and Sergio Ferreira and Salvador Moncada demonstrated that prostaglandin release was blocked from the perfused, isolated dog spleen [49]. These findings provided a simple explanation for the analgesic, anti-inflammatory, antipyretic and antiplatelet therapeutic effects of aspirin [50].

For his work on aspirin, Vane was awarded the Nobel Prize for Physiology and Medicine in 1982.

Conclusions

For most of the 20th century, aspirin has been the most widely used drug in the world, initially for its antipyretic and antirheumatic properties and then for its antithrombotic effects. If introduced into medical research today, it is unlikely that it would progress far beyond the laboratory and would almost certainly be denied a product licence by government regulatory authorities. Yet in America alone, it has been estimated that more than 20 000 million aspirin tablets (100 tablets per citizen) are consumed annually [6], with more than one in four healthy middle-aged men and women reporting taking aspirin regularly to reduce risk of vascular events [51]. In the UK, recent audits have revealed that almost two-thirds of pharmacy sales of aspirin are now used for cardiovascular indications. Of these, one in five is being taken without the family doctor's knowledge [52]. Few other drugs have had such a significant impact on morbidity and mortality or look likely to remain so frequently dispensed well into the next century.

KEYPOINTS

- Salicylate-containing plants have been used since classical times for pain relief and to treat rheumatism. In 1763, an Oxfordshire clergyman, Edward Stone, published a description of the successful treatment with extracts of willow bark of 50 cases of fever and agues.
- Acetylsalicylic acid was first synthesized by Charles Gerhardt in the 1850s, but it was Felix Hoffmann, an employee of the Bayer Chemical Company in Germany, who is credited with first realizing its clinical potential as an antiarthritic and antipyretic.
- The trade name, aspirin, was derived from the plant genus *Spiraea*, a natural source of salicylic acid. It was registered in 1899 and aspirin rapidly became one of the most widely used medications in the world.

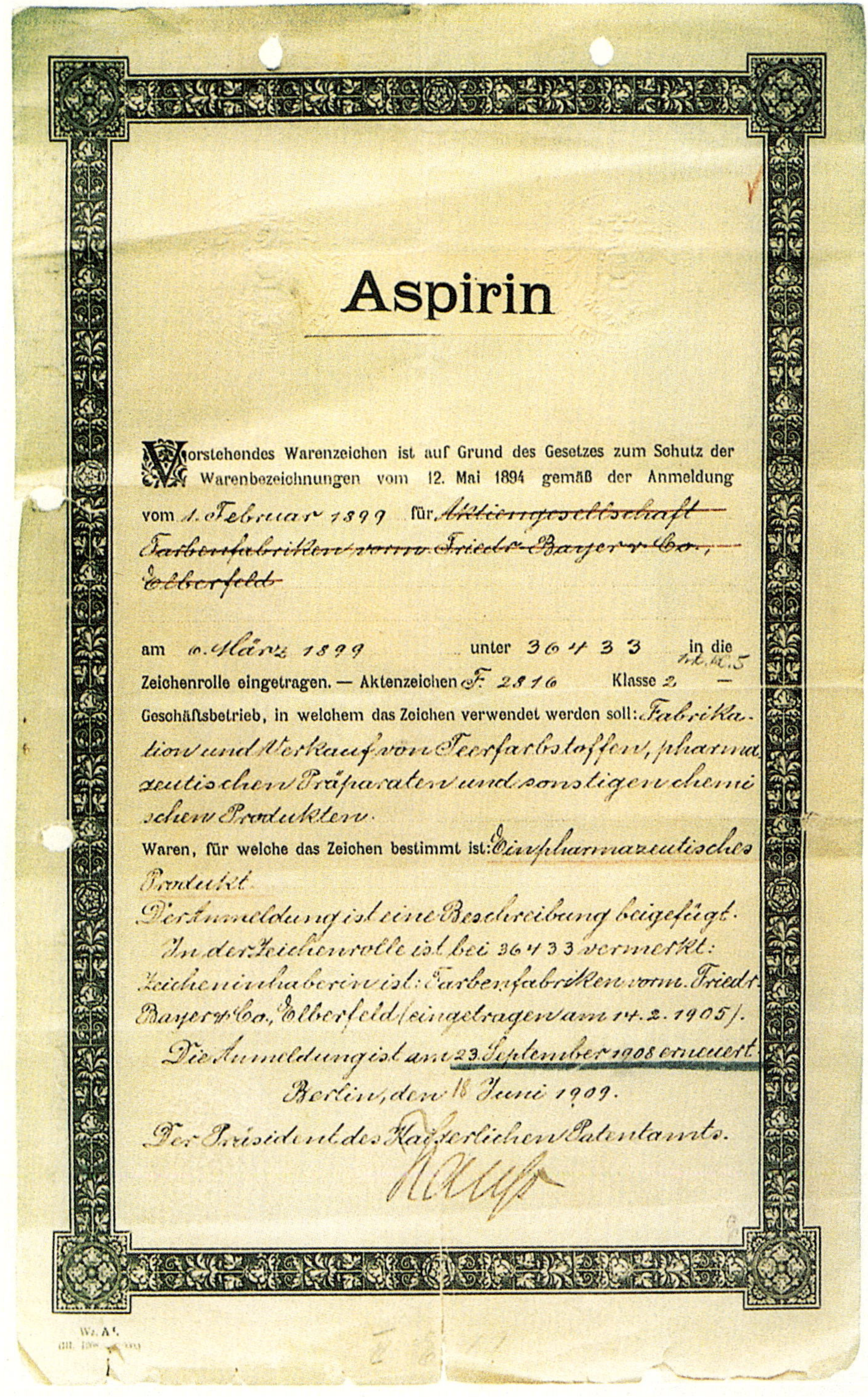

Aspirin

Vorstehendes Warenzeichen ist auf Grund des Gesetzes zum Schutz der Warenbezeichnungen vom 12. Mai 1894 gemäß der Anmeldung vom 1. Februar 1899 für ~~Aktiengesellschaft Farbenfabriken vorm. Friedr. Bayer & Co., Elberfeld~~

am 6. März 1899 unter 36433 in die Zeichenrolle eingetragen. — Aktenzeichen F. 2816 Klasse 2 —

Geschäftsbetrieb, in welchem das Zeichen verwendet werden soll: Fabrikation und Verkauf von Teerfarbstoffen, pharmazeutischen Präparaten und sonstigen chemischen Produkten.

Waren, für welche das Zeichen bestimmt ist: Ein pharmazeutisches Produkt.

Der Anmeldung ist eine Beschreibung beigefügt.

In der Zeichenrolle ist bei 36433 vermerkt: Zeicheninhaberin ist: Farbenfabriken vorm. Friedr. Bayer & Co., Elberfeld (eingetragen am 14.2.1905).

Die Anmeldung ist am 23. September 1908 erneuert.

Berlin, den 18 Juni 1909.

Der Präsident des Kaiserlichen Patentamts.

Plate 1 Certificate of registration with the German Imperial Patent Office, Berlin, on February 1st 1899 (provided by Bayer AG).

Plate 2 Felix Hoffmann, credited with rediscovering the synthesis of acetylsalicylic acid (Bayer AG).

Plates 3 and 4 Early commercial preparations of aspirin (provided by Bayer AG).

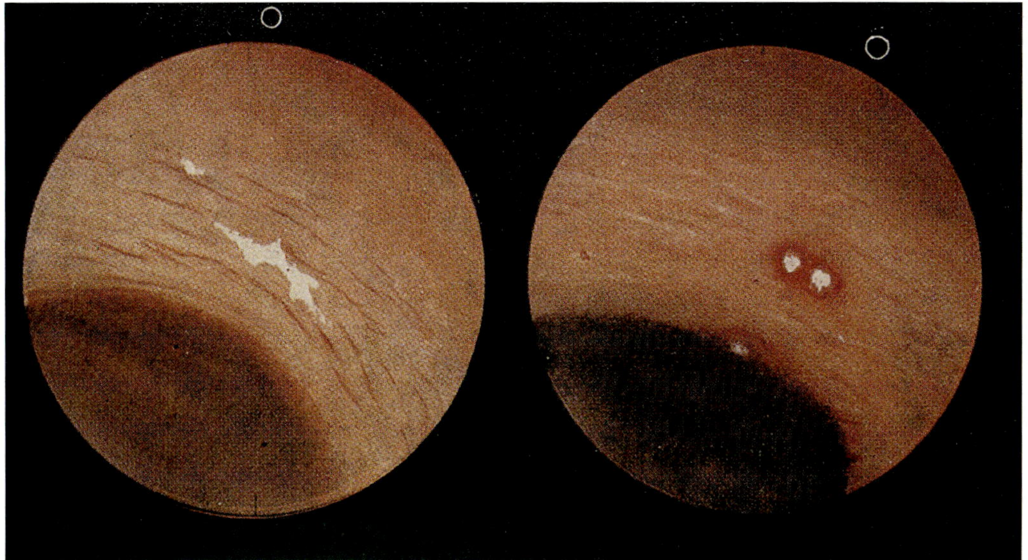

Plate 5 Illustration from Douthwaite and Lintott's paper in The *Lancet* of November 26th 1938, showing the effect on the gastric mucosa of I. barium sulphate (as control, causing no reaction) and II. aspirin (surrounding hyperaemia indicating a localized inflammatory response) (from reference [32], with permission of the Lancet Limited).

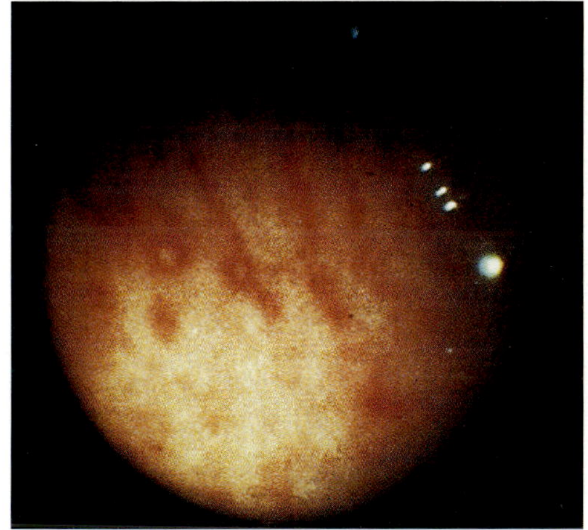

Plate 6 Gastric erosions and punctate and linear haemorrhages associated with aspirin therapy.

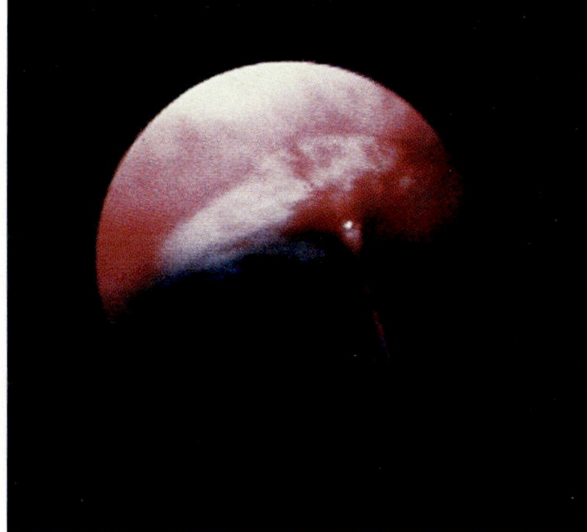

Plate 7 Gastric ulcer with active arterial bleeding.

Plates 8 Poster advertising aspirin from Germany (reproduced with permission from Bayer AG).

Plate 9 Poster advertising aspirin from Spain (reproduced with permission from Bayer AG).

Plate 10 Poster advertising aspirin from Spain (reproduced with permission from Bayer AG).

Plate 11 Poster advertising aspirin from Italy (reproduced with permission from Bayer AG).

Plate 12 Poster advertising aspirin from Germany (reproduced with permission from Bayer AG).

Plate 13 Poster advertising aspirin from China (reproduced with permission from Bayer AG).

Plate 14 Poster advertising aspirin from Turkey (reproduced with permission from Bayer AG).

Plate 15 Poster advertising aspirin from France (reproduced with permission from Bayer AG).

- In the early 1970s, Nobel prize-winning research by Sir John Vane demonstrated that aspirin blocks prostaglandin synthesis and provided a scientific rationale for its analgesic, anti-inflammatory and antipyretic uses and encouraged interest in its antithrombotic properties.
- In the last quarter century, some of the largest clinical trials ever conducted, together with pioneering statistical overviews of published evidence, have firmly established the use of low-dose aspirin for the treatment and prevention of most forms of occlusive cardiovascular disease.

REFERENCES

1. Vane JR, Flower RJ, Botting RM. History of aspirin and its mechanism of action. Stroke 1990; 21(suppl IV): IV-12–IV-23.
2. Friend DG. Aspirin: the unique drug. Arch Surg 1974; 108: 765–769.
3. Lovell R. Lord Moran's prescriptions for Churchill. Br Med J 1995; 310: 1537–1538.
4. Gross M, Greenberg LA. The Salicylates: A Critical Bibliographic Review. New Haven, Conn: Hillhouse Press, 1948: 1–8.
5. Fairley P. The Conquest of Pain. London: Michael Joseph, 1978.
6. Mann CC, Plummer ML. The Aspirin Wars: Money, Medicine and 100 Years of Rampant Competition. New York, NY: Alfred A Knopf Inc, 1991.
7. Mueller RL, Scheidt S. History of drugs for thrombotic disease. Discovery, development, and directions for the future. Circulation 1994; 89: 432–449.
8. Rodnan GP, Benedek TG. The early history of antirheumatic drugs. Arthritis Rheum 1970; 13: 145–165.
9. Rainsford KD. Aspirin and the Salicylates. London: Butterworths, 1984.
10. MacLagan T. The willow as a remedy for acute rheumatism. Lancet 1876; 110: 910.
11. Aronson SM. The miraculous willow tree. Rhode Island Med 1994; 77: 159–161.
12. Stone E. An account of the success of the bark of the willow in the cure of agues. Philosoph Trans R Soc London 1763; 53: 195–200.
13. Meades E. The History of Chipping Norton. Chipping Norton: Bodkin Bookshop, 1984.
14. MacLagan TJ. The treatment of acute rheumatism by salicin. Lancet 1876; 1: 342–343, 383–384.
15. Editorial. Quinine, salicine and allied drugs. Br Med J 1877; 1: 117.
16. Campbell H. The Salicylic Treatment of Gout, Rheumatic Gout, Neuralgia and Diabetes, 2nd edn. London: H. Renshaw, 1879.
17. Sneader W. Drug Discovery. The Evolution of Modern Medicines. Chichester: John Wiley and Sons, 1985.

18. Riess L. Ueber die innerliche Anwendung der Salicylsaure. Berl Klin Wochnschr 1875; 12: 673–676.
19. Stricker F. Uber die resultate der behandlung der polyarthritis rheumatica mit salicylsaure. Berl Klin Wochenschr 1876; 13: 1–2, 15–16, 99-103.
20. See G. Etudes sur l'acide salicylique et les salicylates; traitement du rhumatisme aigu et chronique de la goutte, et de diverses affections du systeme nerveux sensitif par les salicylates. Bull Acad Med Paris 1877; 6: 689, 897.
21. Editorial. Special correspondence: Paris. M. See on salicylic acid. Br Med J 1877; 2: 865.
22. Schwarz A. Beethoven's renal disease based on his autopsy: a case of papillary necrosis. Am J Kidney Dis 1993; 21: 643–652.
23. Sharp G. The history of the salicylic compounds and of salicin. Pharm J 1915; 94: 857.
24. Collier HOJ. Aspirin. Scientific American 1963; 209: 108.
25. Weissmann G. Aspirin. Scientific American 1991; 264: 58–64.
26. Krantz JC, White JM. Aspirin: four score and more. South Med J 1980; 73: 1630–1634.
27. Dreser H. Pharmakologisches uber Aspirin (Acetylsalicylsaure). Pfluger's Archiv Anat Physiol 1899; 76: 306–318.
28. Witthaeuer K. Aspirin, ein neues Salicyl-preparat. Die Heilkunde 1899; 3: 396.
29. Wohlgemut J. Uber Aspirin (Acetylsalicylsaure). Therapeutische Montatshefte (Halbmh) 1899; 13: 276–278.
30. Floeckinger FC. An experimental study of aspirin, a new salicylic acid preparation. Med News 1899; 75: 645.
31. Burnet J. The therapeutics of aspirin and mesotan. Lancet 1905; 168: 1193–1196.
32. Douthwaite AH, Lintott GAM. Gastroscopic observation of the effect of aspirin and certain other substances on the stomach. Lancet 1938; ii: 1222–1225.
33. Hurst A, Lintott GAM. Aspirin haematemesis. Lancet 1939; ii: 843.
34. Link KP, Overman RS, Sullivan WR, Huebner CF, Scheel LD. Studies of the hemorrhagic sweet clover disease. XI. Hypoprothrombinaemia in the rat induced by salicylic acid. J Biol Chem 1943; 147: 463–474.
35. Quick AJ. Bleeding-time after aspirin ingestion. Lancet 1968; i: 50.
36. Gibson PC. Aspirin in the treatment of vascular diseases. Lancet 1949; ii: 1172–1174.
37. Craven LL. Acetylsalicylic acid, possible preventive of coronary thrombosis. Ann West Med Surg 1950; 4: 95.
38. Craven LL. Experiences with aspirin (acetylsalicylic acid) in the non-specific prophylaxis of coronary thrombosis. Mississippi Valley Med J1953; 75: 38–40.
39. Craven LL. Prevention of coronary and cerebral thrombosis. Mississippi Valley Med J 1956; 78: 213–215.
40. Anonymous. Aspirin and the blood. Lancet 1966; 2: 787.

41. Weiss HJ, Aledort LM, Kochwa S. The effect of salicylates on the hemostatic properties of platelets in man. J Clin Invest 1968; 47: 2169–2180.
42. O'Brien JR. Effects of salicylates on human platelets. Lancet 1968; i: 779-783.
43. Elwood PC, Cochrane AL, Burr ML *et al.* A randomised controlled trial of acetyl salicylic acid in the secondary prevention of mortality from myocardial infarction. Br Med J 1974; 1: 436–440.
44. ISIS-2 (Second International Study of Infarct Survival) Collaborative Group. Randomised trial of intravenous streptokinase, oral aspirin, both, or neither among 17,187 cases of suspected acute myocardial infarction. ISIS-2. Lancet 1988; 2: 349–360.
45. ASPIRE Steering Group. A British Cardiac Society survey of the potential for the secondary prevention of coronary disease: ASPIRE (Action on Secondary Prevention through Intervention to Reduce Events). Heart 1996; 75: 334–342.
46. Antiplatelet Trialists' Collaboration. Collaborative overview of randomised controlled trials of antiplatelet therapy. I: Prevention of death, myocardial infarction, and stroke by prolonged antiplatelet therapy in various categories of patients. Br Med J 1994; 308: 81–106.
47. Vane JR. Inhibition of prostaglandin synthesis as a mechanism of action for aspirin-like drugs. Nature 1971; 231: 232–235.
48. Smith JB, Willis AL. Aspirin selectively inhibits prostaglandin production in human platelets. Nature 1971; 231: 235–237.
49. Ferreira SH, Moncada S, Vane JR. Indomethacin and aspirin abolish prostaglandin release from spleen. Nature 1971; 231: 237–239.
50. Editorial. Prostaglandins, aspirin and analgesia. Lancet 1973; i: 979.
51. Shahar E, Folsom AR, Romm FJ *et al.* Patterns of aspirin use in middle-aged adults: the Atherosclerosis Risk in Communities (ARIC) Study. Am Heart J 1996; 131: 915–922.
52. Anonymous. Aspirin: public get the message. Prescriber 1997; 8: 15.

Pharmacokinetics and pharmaceutical considerations

Introduction

A basic understanding of the chemical and physical properties of a drug and its mode of absorption, distribution and metabolism within the body will help to explain its therapeutic activity and associated adverse effects. This will ensure that the drug can be used optimally and that unnecessary problems are avoided. In the case of aspirin, the pharmacokinetics are complex [1,2], with considerable intra- and intersubject variability, even at low dosage [3].

Physicochemical properties

Aspirin is a moderately weak acid with a pK_a of 3.5. It has relatively high lipophilicity and poor aqueous solubility, especially at low pH. In its natural state, aspirin exists as a white, crystalline powder, which is odourless or with a faint smell of acetic acid [4]. It is stable in dry air but when exposed to moisture it rapidly undergoes hydrolysis. As well as humidity, the rate of decomposition is dependent on temperature and pH [5]. At pH 2.25 its solubility is reported as 3.4 g/l whereas at pH 7 it is 8220 g/l [6]. Aspirin tablets should therefore be stored in airtight containers, at temperatures not exceeding 25°C [7]. The most effective means of preventing decomposition of aspirin suppositories is refrigeration [8]. Aspirin solutions should always be freshly prepared and special care is needed when collecting and processing biological fluids for measurement of drug levels.

MEASUREMENT IN BIOLOGICAL FLUIDS

To avoid rapid hydrolysis, blood samples for aspirin assay should be taken into fluoride tubes and separated and frozen immediately [9]. Analysis should be completed promptly as, even when stored in dry ice, the hydrolysis half life is 24 days [10]. In practice, it is usually salicylate levels which are measured and it is then important to consider free (pharmacologically active) levels, as well as those of total drug.

Various methods are available for measuring plasma, serum, salivary and urinary salicylate levels. Historically, the spectrophotometric method developed by Trinder [11] was widely used. This was based on the chelation of iron by salicylic acid to yield a purple-coloured complex whose absorbance was then measured by a spectrophotometer. The specificity and sensitivity of this method were problematic as other natural phenols and amino acids were absorbed in the same region. Alternative analytical methods were subsequently developed, using gas–liquid chromatography, spectrofluorimetry and even thin-layer chromatography. All have now been superseded by high-performance liquid chromatographic (HPLC) methods, which quickly and reliably measure aspirin, salicylic acid and all its metabolites [12,13].

In the clinical situation, measurement of plasma salicylate concentrations is now mainly used to guide management when aspirin has been taken in overdose. In therapeutic dosage, plasma monitoring may help to avoid side effects and can be used to check compliance. Blood samples are usually best taken immediately preceding the next dose of drug, to provide 'trough' plasma levels.

Salivary concentrations of salicylic acid are dependent on the pH at the site of saliva production, but are generally proportional to plasma concentrations of free (unbound) salicylic acid and may sometimes be useful for monitoring when repeated blood sampling is difficult or undesirable [14].

Pharmacokinetics

ABSORPTION

When taken by mouth, aspirin is rapidly absorbed, mainly intact, partly from the stomach but chiefly from the upper small intestine. Salicylates are also readily absorbed in the large bowel and colon. Absorption from the oral mucosa is slow, probably due to the limitation of the available surface area [1]. It is increased if tablets are chewed. Absorption at all sites occurs by

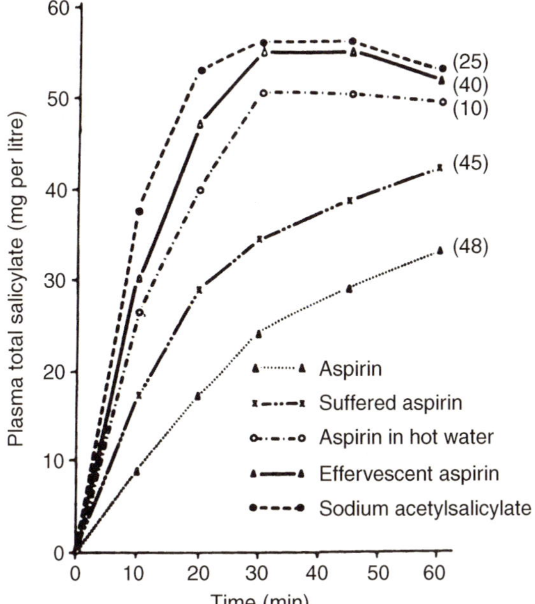

Fig. 2.1
Plasma salicylate concentrations after oral ingestion of various preparations of aspirin (640 mg equivalent). Figures in brackets refer to number of patients in each group (from reference [1] with permission).

passive diffusion of primarily non-ionized drug across the mucosal membrane. Spontaneous hydrolysis of aspirin to salicylic acid is minimal and though there is some hydrolysis by carboxyl-esterases in the gut wall, most absorption is of the parent drug rather than of free salicylate [15,16].

Aspirin administered in aqueous solution is rapidly absorbed, for example from effervescent preparations (e.g. Alka-Seltzer) [17] (Figure 2.1). The rate of absorption of aspirin from solid dosage forms is determined by the disintegration time and dissolution characteristics of the particular pharmaceutical formulation administered (tablets, capsules, powder or dispersion, buffered or enteric coated) [18,19]. Differences in starch content, granule size, compression pressure and type and concentration of tablet lubricant will all markedly affect the dissolution rate [1].

Other factors which will affect aspirin absorption are the pH of the mucosal absorptive area, the gastric emptying time (itself increased at higher pH), the presence of food in the stomach, posture and activity at time of ingestion, concurrent drugs and any disease state associated with altered gastrointestinal transit time [2]. Age has no influence on absorption [20–22] (Table 2.1).

Aspirin at pH below 3.5 (pK_a or isoelectric point) is non-ionized and lipophilic. As the pH rises, the amount of salicylate absorbed from the gut decreases. Thus the generally low pH of the stomach should theoretically favour absorption there, rather than in the higher pH environment of the small bowel. However, as pH increases in the duodenum and jejunum,

Table 2.1
Factors influencing rate of aspirin absorption

- Rate of tablet dissolution
- Gastric and small intestinal pH
- Gastric emptying
- Presence of food
- Other drugs
- Posture
- Exercise
- Diseases associated with altered GI transit

aspirin becomes more ionized and so more water soluble. This will favour dissolution of aspirin tablets, which is maximal at pH 8. More importantly, the small intestine offers a much larger absorptive surface area than the stomach. The net effect is that most aspirin absorption occurs in the small bowel [1].

After oral ingestion of a single dose of an aqueous solution of aspirin, the absorption half life is between four and 16 minutes and follows first-order kinetics [16]. At least two-thirds of the dose reaches the systemic circulation unhydrolysed, the remainder presumed to be metabolized by esterases within the gut wall, liver or plasma. Peak concentrations of aspirin taken in therapeutic dosage are found in plasma within 30 minutes of ingestion of soluble aspirin [23], within two hours of conventional tablets and within about 4–6 hours of enteric-coated preparations [24]. Plasma aspirin levels then rapidly decline as salicylic acid levels rise.

If aspirin is taken in therapeutic overdose, either intentionally or by accident, absorption may occur more slowly due to impaired dispersion and the inhibitory effect of aspirin on gastric emptying. Peak plasma salicylate concentrations may then continue to rise for up to 24 hours [25].

DISTRIBUTION

After absorption, salicylate is distributed widely throughout most body tissues, with highest concentrations occurring in the kidney, liver, heart and lungs. It penetrates well into synovial fluid, peritoneal fluid and saliva, but only slightly into bile, sweat and inflammatory exudates and not at all into gastric juice [26]. At therapeutic concentrations the apparent volume of distribution in adults ranges from 9.6 to 12.7 litres [27] or about 170 ml/kg body weight [28], though there is marked intersubject variation. Some studies have suggested that the volume of distribution is increased in elderly patients [22,29], whereas others have shown no change [20,30]. In practice, any age-related effects are insignificant.

At anti-inflammatory concentrations in patients with rheumatoid arthritis or osteoarthritis, salicylic acid is 80–90% bound to plasma proteins, particularly albumin [31]. There are at least two primary binding sites of salicylate to albumin and a number of secondary sites [32] and salicylate competes for these with a variety of compounds, including thyroid hormone, bilirubin, uric acid and drugs such as penicillin, phenytoin and other non-steroidal anti-inflammatory drugs. Conditions such as renal failure [33] and pregnancy and neonatal age [34] are associated with decreased protein binding of salicylate. At high plasma salicylate concentrations, binding sites become saturated and the free fraction thus increases. At high dosage, therefore, distribution volume increases by as much as threefold, to 500 ml/kg. Hypoalbuminaemia, as often occurs with acute inflammatory disease, will also result in proportionately higher free salicylate levels. Aspirin itself appears to bind less strongly than salicylate to albumin, but in doing so permanently acetylates the protein molecule by reaction with the amino group of the 199 lysine residue [35]. This may result in altered binding of other compounds.

Distribution throughout the body is mainly by pH-dependent passive diffusion [36]. There is some evidence from *in vitro* studies that it may be actively transported in the cerebrospinal fluid [37]. Total aspirin and salicylic acid concentrations in synovial fluid are lower than in plasma [38], probably because of less protein binding [39], but the unbound concentrations of salicylate in plasma and synovial fluid tend to equalize with chronic, high-dose treatment. Peak concentrations of aspirin in synovial fluid also occur much later than in plasma but persist for longer, indicating slow transport into and out of the joint [40].

Salicylate is readily transferred across the placenta and neonatal concentrations will be higher than concurrent maternal concentrations because of immature biotransformation and excretory pathways [34]. There is also ready distribution of salicylate into breast milk, but the amount transferred after maternal ingestion of a single dose of aspirin is small. A nursing mother taking regular large doses of aspirin, however, may expose an infant to more significant levels of salicylate [41].

METABOLISM

Aspirin is rapidly hydrolysed in the body to salicylic acid by non-specific carboxyl-esterases which are present in many body tissues [42]. There is an inverse correlation between aspirin esterase activity and the plasma half life of salicylate, indicating that it is this enzyme which largely controls the rate-limiting conversion of aspirin to salicylate. Any lowering of enzyme activity will increase the proportion of aspirin in plasma and vice versa [43]. The activity of aspirin esterase is significantly influenced by the presence of medical illness and physical frailty, though not by healthy ageing [44,45]. It has been suggested that patients prone to duodenal ulceration have subnormal gastric aspirin esterase activity, whereas those with gastric ulceration or gastric cancer have normal enzyme activity [43]. Inter-racial differences have also been reported, but it is uncertain if these are true genetic differences or merely relate to differences in nutritional status [46]. There is some evidence of diurnal variation in the pharmacokinetics of aspirin [47].

The process of hydrolysis starts during absorption. Up to 35% of the hydrolysis of orally administered aspirin normally occurs in the gastrointestinal mucosa [16], particularly in the mucosal cells of the gastric fundus [48]. Hydrolysis continues on first pass through the liver (accounting for another 15–30% of the ingested dose) and by esterases in serum and red blood cells.

The *in vitro* half life of aspirin in plasma is about 15–20 minutes [16] and the elimination half life after intravenous dosing is similar [49]. After oral dosing aspirin is detectable in plasma for several minutes before there is any measurable salicylic acid and remains the dominant form of the drug in the plasma for up to 20 minutes after ingestion. It then disappears rapidly [2]. Thus the plasma aspirin concentration is very dependent on the rate of absorption and is never high, rarely exceeding 20 mg/ml after therapeutic doses. With repeated dosing it is salicylic acid but not aspirin which will accumulate until steady state is reached.

The biotransformation of salicylic acid occurs chiefly in the hepatic endoplasmic reticulum and mitochondria (Figure 2.2). The three chief metabolic products are salicyluric acid (the glycine conjugate) and salicylic acyl and phenolic glucuronides [50]. A small amount is oxidized to gentisic acid and its glycine conjugate, gentisuric acid [51]. Some studies have suggested sex differences in salicylate metabolism, with intrinsically lower metabolic activity in women [52,53]. However, any differences were small in magnitude and probably of no clinical significance.

The elimination pathways to salicyluric acid and to salicyl phenolic glucuronide are saturable, following Michaelis–Menten kinetics, whereas the others all follow first-order kinetics [54,55]. Thus elimination is only linear after the amount remaining in the body has decreased to about 400 mg equivalent or less [1]. The plasma elimination half life of salicylic acid is therefore dose dependent (Figure 2.3), ranging from about 2–3 hours for low doses of aspirin to 12 hours or more when anti-inflammatory doses are given and up to 30 hours when aspirin is taken in overdose. Small

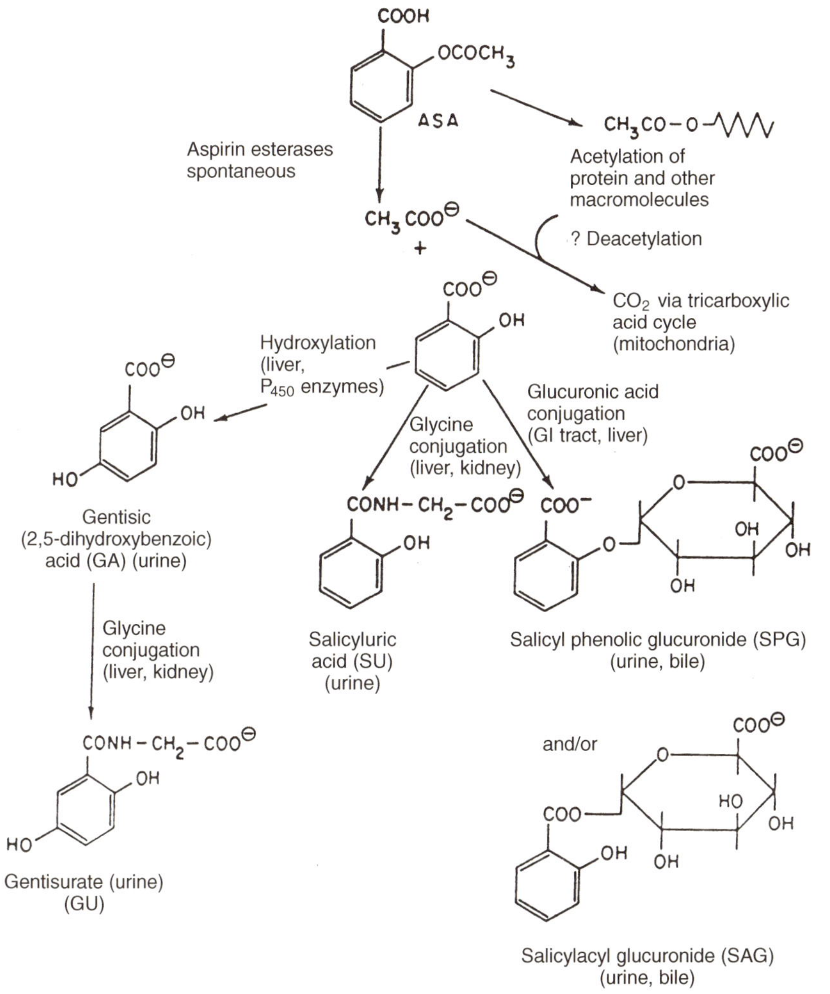

Fig. 2.2
Structure and metabolism of aspirin (ASA) and salicylic acid in man (from reference [43] with permission).

increases in aspirin dosage may therefore result in disproportionate increases in plasma levels of salicylate. A study in rheumatoid arthritis patients demonstrated that a 50% increase in the daily dose of aspirin (from 65 to 100 mg/kg) resulted in a 300% rise in serum salicylate concentration [56]. The time required to reach steady state will also increase with increasing daily dose, from about 48 hours when 1.5 g/day is taken to a week or more when the dose is doubled. With chronic use, salicylate may induce its own metabolism by increasing salicyluric acid production [57] Certainly plasma salicylic acid levels decline significantly with chronic administration in normal volunteer studies [58] and in patients with rheumatoid arthritis, although therapeutic response seemed not to be affected [59].

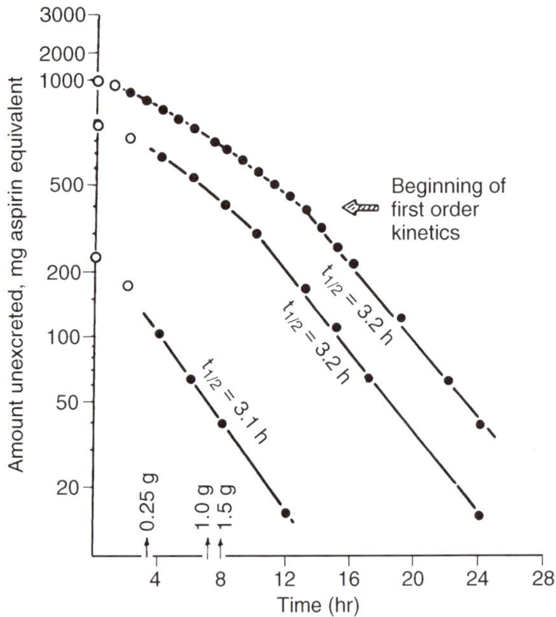

Fig. 2.3
Elimination of salicylate by a 23-year-old normal adult male as a function of dose (250 mg, 1000 mg and 1500 mg). Vertical arrows on the time axis indicate the time necessary to eliminate 50% of the dose (from reference [103] with permission).

EXCRETION

Salicylic acid is removed from the body by renal elimination of both free (unchanged) and metabolized drug and depends on urinary pH, flow rate and presence of organic acids. Minor extrarenal pathways account for less than 10% of the total dose. Free salicylic acid, its glucuronide conjugates and gentisic acid all diffuse freely across the glomerular membrane and are also actively secreted by the proximal tubule. Only the non-ionized form of salicylic acid undergoes renal tubular reabsorption and is therefore pH dependent [2].

After oral dosing of a standard 300 mg aspirin tablet about 75% of salicylate in the urine is salicyluric acid, 10% the phenolic glucuronide, 5% the acyl glucuronide and less than 1% is gentisic acid [1]. With a 3 g dose of aspirin, 50% of the dose is excreted as salicyluric acid, 20% as phenolic glucuronide, 10% as the acyl glucuronide and 3% as gentisuric acid [60]. Renal clearance of salicyluric acid decreases significantly with age and correlates significantly with creatinine clearance [29].

Free salicylic acid generally makes up about 10–15% of urinary salicylate, but this depends on dose and on urinary pH and can be very variable (Figure 2.4). In alkaline urine, 30% or more of a therapeutic dose may be excreted as free salicylic acid, whereas in acid urine it may be less than 2% [4]. Changes in urinary pH due to antacid therapy will cause lower steady-state plasma salicylate levels [61,62] and patients taking therapeutic doses of

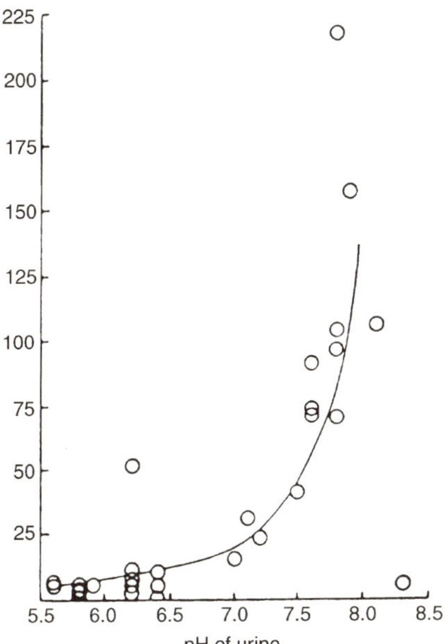

Fig. 2.4
Relationship between urinary pH and renal clearance of free salicylate (from reference [104] with permission).

aspirin may become intoxicated on withdrawal of a concurrent antacid. In general, however, the effect of urinary pH is only important at high salicylate concentrations, when a greater proportion will be excreted unchanged [63]. This is of particular importance in managing patients overdosed with aspirin.

Renal excretion of salicylic acid is also sensitive to urinary flow rate, with high rates decreasing tubular reabsorption, whereas the opposite will be true in oliguria. Recovery of salicylic acid is negatively correlated with that of salicyluric acid and when plasma salicylate levels are very high, salicylic acid may account for 60% or more of total urinary metabolites [55].

Which preparation?

Many varied pharmaceutical preparations are available worldwide [7]. In Britain (Table 2.2), aspirin is available as acetylsalicylic acid tablets, as dispersible (soluble) tablets and as effervescent tablets, in doses from 75 mg to 500 mg. It is also compounded with caffeine and with codeine phosphate (Co-codaprin), in proportions by weight of 50 parts aspirin to one part codeine. Enteric-coated and modified-release preparations are also marketed. Altogether, over 40 different preparations are available, either containing aspirin alone or combined with other ingredients. These range from the familiar Alka-Seltzer and Beechams Powders to the less familiar

Name	Dose (mg)	Formulation	Other main constituents
Non-proprietary preparations			
Aspirin	75,300	tab, dis	
Aspirin suppositories	150,300	sup	
Co-codaprin tablets	400	tab, dis	Codeine
Proprietary preparations			
Angettes 75	75	tab	
Aspro	320,500	tab, dis	
Bayer Aspirin	300	tab	
Beechams Lemon Tablets	300	tab	
Caprin	75,300	e/c, f/c	
Disprin CV	100,300	tab, m/r	
Disprin Direct	300	dis	
Nu-seals Aspirin	75,300	e/c	
PostMI 75	75	e/c	
Multi-ingredient preparations			
Alka-Seltzer	324	dis	Citrate, bicarbonate
Alka-Seltzer XS	267	dis	Paracetamol, caffeine
Anadin	325	e/c, tab, dis	Caffeine
Anadin Extra	300	tab, dis	Paracetamol, caffeine
Anadin Maximum	500	cap	Caffeine
Anadin–paracetamol	500	tab, cap	Paracetamol
Askit	530	dis	Caffeine
Aspav	500	dis	Mixed opium alkaloids
Beechams Powders	600	dis	Caffeine
Beechams Tablets	300	tab	Caffeine
Codis	500	tab	Codeine
Disprin Extra	300	tab	Paracetamol
Doloxene Compound	375	cap	Dextropropoxyphene, caffeine
Equagesic	250	tab	Meprobamate, ethoheptazine
Femigraine	500	tab	Cyclizine
Fynnon Calcium Aspirin	500	tab	Calcium carbonate
Mrs Cullen's Powders	62	sol	Caffeine
Nurse Syke's Powders	165	tab, dis	Caffeine, paracetamol
Phensic	325	tab, dis	Caffeine
Powerin	300	tab	Caffeine, paracetamol
Robaxisal Forte	325	tab	Methocarbamol
Toptabs	350	tab	Caffeine
Veganin	250	tab	Paracetamol, codeine

Table 2.2
Preparations containing aspirin available in the UK in 1997 tab, tablet; dis, dispersible; sup, suppository; e/c, enteric coated; cap, capsule; f, film coated; sol, soluble; m/r, modified release

Nurse Sykes's Powders and Mrs Cullen's Powders. In America and many other countries, numerous buffered aspirin preparations and capsule and suppository formulations are also available, as well as flavoured and sweetened preparations [4]. A commercial parenteral preparation is not available, but the lysine salt of aspirin has occasionally been used therapeutically [64] and the *N*-methylglucamine salt has been used in pharmacological studies [49].

The major factor which should influence the choice of preparation to be used is the speed of onset of action which is required. Other issues to be considered are the convenience and practicality of taking a solid or liquid preparation, concern about problems of gastric intolerance and cost. In Britain, pack size restrictions have recently been introduced with the aim of limiting the availability of aspirin in the home which might be used in deliberate or accidental overdose. The restrictions do not apply to effervescent forms, granules or powders, but limit the number of aspirin tablets which can be sold over the counter to 16 tablets through general sales outlets such as supermarkets and 32 tablets through pharmacists. Above this quantity, a prescription is required [65].

SOLUBLE AND EFFERVESCENT PREPARATIONS

Aspirin is most rapidly absorbed when administered in aqueous solution. This will also reduce the risk of solid particles of aspirin sitting on the gastric mucosa and producing localized inflammatory reactions by acting as a source of concentrated acetylsalicylic acid. In general, the bioavailability of effervescent (e.g. Alka-Seltzer) and soluble aspirin preparations is greater than conventional tablets and the absorption profiles are similar [66,67]. Peak plasma salicylic acid levels are attained after about 30 minutes on average and this may be therapeutically useful when a rapid onset of action is required, for example after onset of symptoms of possible acute myocardial infarction or in treatment of acute migraine.

Aqueous solutions of aspirin must be treated with care, as aspirin rapidly undergoes hydrolysis to acetic and salicylic acids [68]. A freshly prepared solution at neutral pH and stored at room temperature for four hours will undergo 2% hydrolysis to salicylic acid, whereas after 24 hours 10% or more will have been hydrolysed [69].

ENTERIC-COATED PREPARATIONS

Enteric-coated aspirin has an outer coating of cellulose or silicon or other substance which does not dissolve until after the tablet has passed through the stomach and reached the more alkaline or neutral environment of the duodenum. Thus the gastric mucosa is theoretically protected from irritation by locally released aspirin, with the assumption that there will be less upper gastrointestinal bleeding. In general, absorption is slower from enteric-coated preparations, with about 90% of the total dose absorbed [70], but otherwise the metabolic profile and pharmacological effects of standard and enteric-coated aspirin preparations are similar [71,72].

Absorption from enteric-coated preparations is very dependent on gastrointestinal motility. If gastric emptying is prolonged, absorption will be delayed and may be incomplete [73]. An early *in vivo* study suggested that absorption was especially slow if tablets settled into the fundus of the stomach, where there is relatively little movement [74]. Furthermore, there can be considerable differences between different enteric-coated preparations. A preparation containing enteric-coated aspirin granules was shown to have more rapid and reliable absorption in the small intestine than a conventional enteric-coated tablet [75].

Studies in healthy volunteers comparing standard and enteric-coated aspirin have appeared to confirm the theoretical advantages of the enteric-coated preparations, with less evidence of gastric mucosal damage and microbleeding assessed endoscopically [76–78] and less blood loss measured by estimation of haemoglobin in gastric washings [78] and by measurement of radiochromium-labelled erythrocytes in three-day stool collections [79].

Clinical studies, however, have been less supportive of protective effects of enteric coating, at least when low daily doses of 325 mg or less of aspirin are being used. Thus in a randomized controlled trial of enteric-coated aspirin 100 mg daily versus placebo given to 400 healthy elderly patients for 12 months, haematemesis and/or melaena occurred in six patients on active medication and in no placebo patients [80]. Similarly, a case control study of over 500 endoscopically confirmed cases of upper gastrointestinal bleeding reported relative risk of bleeding to be 2.6 for standard aspirin and 2.7 for enteric-coated aspirin at average daily doses of 325 mg or less. At higher doses the relative risk was 5.8 for standard aspirin, but there were insufficient data to evaluate enteric-coated aspirin [81]. It is possible that at higher dosage, enteric coating may offer some clinically useful protection, at least during very short-term administration, when there is very little or no contact with the gastric mucosa [82,83].

The delay after ingestion until release of the active ingredient in the duodenum suggests that enteric-coated preparations are not appropriate for quick relief of acute pain or for rapid antithrombotic action after acute myocardial infarction or unstable angina. If chewed, absorption will be more rapid but the theoretical benefits of enteric coating will be lost. The simultaneous ingestion of antacids may also lead to premature drug release.

BUFFERED PREPARATIONS

The combination in the same tablet of aspirin with buffering agents, such as calcium carbonate, sodium bicarbonate, magnesium oxide and magnesium carbonate, was developed with the expectation that gastric mucosal damage would be reduced. It was assumed that the buffering agent would lower the pH in the immediate environment of the dissolving aspirin particles, increasing aspirin solubility, reducing contact time with the gastric mucosa and so limiting risk of gastrointestinal bleeding [84,85]. A study of the effects of 11 different buffering agents found that most resulted in more rapid dissolution of aspirin from tablets [86]. However, the effect on absorption is dose dependent and it is therefore difficult to generalize. Buffered preparations containing 16 mmol of buffer are likely to be absorbed more rapidly than those containing 32 mmol in the fasted [87] and unfasted state [88]. Total salicylate absorption is not altered.

Unfortunately endoscopic studies have generally failed to confirm any protective effect compared with standard aspirin [85,89]. A case control study of over 500 endoscopically confirmed cases of upper gastrointestinal haemorrhage reported relative risk of bleeding to be 2.6 for standard aspirin and 3.1 for buffered aspirin at average daily doses of 325 mg or less. At doses greater than 325 mg, the relative risk was 5.8 for standard and 7.0 for buffered aspirin. [81]. A large observational study in patients with rheumatoid arthritis taking relatively large doses of aspirin (greater than 650 mg daily) did report a lower toxicity score for buffered (1.10) than for standard aspirin (1.36) [90] and it is possible that it may have some advantages when high doses are required. Certainly some patients claim that they can better tolerate buffered than standard preparations. In general, however, there would seem to be no evidence of clinical advantage over standard aspirin and it is significantly more expensive.

MODIFIED-RELEASE PREPARATIONS

The development of modified-release preparations of aspirin is a relatively new phenomenon. At high doses (3 g or more daily), studies have shown less faecal blood loss (measured by radiochromium-labelled erythrocytes) in healthy volunteers taking modified-release than with standard aspirin [91,92]. A further theoretical benefit of a modified-release preparation is that by delivering low levels of aspirin to the portal circulation, thromboxane A2 is presystemically inhibited in circulating platelets, but system prostacyclin in vascular endothelium is not suppressed [93–95]. The real clinical benefits of these findings are uncertain. The only available preparation in Britain is the low-dose Disprin CV. This consists of aspirin coated with a diffusion polymer which dissolves to gradually release the active compound throughout the gut over four hours. The cost of low-dose prophylaxis with modified-release aspirin is about 30 times that of using generic dispersible tablets.

RECTAL SUPPOSITORY

Rectal suppositories of aspirin are not commercially available in Britain, but have sometimes been used therapeutically when patients could not take their medication orally [65]. Absorption of aspirin from suppositories is slower, often incomplete and generally less reliable than oral administration and in a study in healthy volunteer subjects, complete absorption only occurred after retention of suppositories for over 10 hours [96]. High blood salicylate concentrations are thus difficult to obtain. Standard aspirin tablets should never be administered rectally as they cause local irritation and there are reports of mucosal erosion and anorectal stenosis [97].

TOPICAL PREPARATIONS

A number of topical aspirin preparations have been developed for direct application to the skin, both for prophylactic systemic antithrombotic effect [98] and for local analgesic and anti-inflammatory effect [99,100] and as a possible treatment for histamine-mediated itch [101]. Clinical experience with these preparations is limited and caution is probably appropriate given the strong keratolytic and locally irritant properties of salicylic acid.

Conclusions

In summary, there may be advantages in taking enteric-coated, buffered or modified-release preparations of aspirin if high doses are required. At lower doses, there is little evidence that the risk of major gastrointestinal bleeding is reduced when these formulations are used rather than standard aspirin, which is considerably cheaper [102]. In the acute situation, effervescent aspirin is most rapidly absorbed and therefore most likely to have an early onset of effect.

KEYPOINTS

- Oral aspirin is well absorbed, partly from the stomach but mainly from the small intestine. Rate of absorption is greatly influenced by characteristics of the formulation taken, as well as mucosal pH and gastric emptying time.
- Aspirin is rapidly hydrolysed in the body to salicylic acid, which distributes freely throughout most body tissues and binds to plasma proteins, especially albumin.
- The three chief metabolic products are salicyluric acid (the glycine conjugate) and salicylic acyl and phenolic glucuronides. Due to saturation kinetics, small increases in aspirin dosage may result in disproportionate increases in plasma levels of salicylate.
- Elimination is mainly renal and dependent on the amount of drug in the body, urinary pH, flow rate and the presence of organic acids.
- Many varied pharmaceutical formulations are available worldwide. Standard or dispersible tablets are appropriate for most purposes and act quickly. Enteric-coated, buffered and controlled-release preparations may cause fewer gastrointestinal symptoms and offer advantages to some patients.

REFERENCES

1. Levy G, Leonards JR. Absorption, metabolism, and excretion of salicylates. In: Smith MJH, Smith PK (eds) The Salicylates. A Critical Bibliographic Review. New York: Interscience, 1966: 5–48.
2. Needs CJ, Brooks PM. Clinical pharmacokinetics of the salicylates. Clin Pharmacokinetics 1985; 10: 164–177.
3. Benedek IH, Joshi AS, Pieniaszek HJ, King SY, Kornhauser DM. Variability in the pharmacokinetics and pharmacodynamics of low dose aspirin in healthy male volunteers. J Clin Pharmacol 1995; 35: 1181–1186.
4. Reynolds JEF (ed.). Martindale: The Extra Pharmacopoeia, 31st edn. London: Royal Pharmaceutical Society, 1996: 17–22.
5. Leeson LJ, Mattocks AM. Decomposition of aspirin in the solid state. J Am Pharmaceut (Sci.) 1958; 47: 329–333.
6. Thiessen JJ. Pharmacokinetics of salicylates. In: Barnett HJM, Hirsh J, Mustard JF (eds) Acetylsalicylic Acid. New Uses for an Old Drug. New York: Raven Press, 1982: 49–61.

7. Lund W (ed.). The Pharmaceutical Codex, 12th edn. London: Pharmaceutical Press, 1994: 741–746.
8. Whitworth CW, Luzzi LA, Thompson BB, Jun HW. Stability of aspirin in liquid and semisolid bases. II Effect of fatty additives on stability in a polyethylene glycol base. J Pharmaceut Sci 1973; 62: 1372–1374.
9. Rowland M, Riegelman S. Determination of acetylsalicylic acid in plasma. J Pharmaceut Sci 1967; 56: 717–720.
10. Walter LJ, Biggs DF, Coutts RT. Simultaneous GLC estimation of salicylic acid and aspirin in plasma. J Pharmaceut Sci 1974: 63; 1754–1758.
11. Trinder P. Rapid determination of salicylate in biological fluids. Biochem J 1954; 57: 301–303.
12. Rumble RH, Roberts MS, Wanwimolruk S. Determination of aspirin and its major metabolites by high pressure liquid chromatography with solvent extraction. J Chromatography 1981; 225: 252–260.
13. Peng GW, Gadalla MA, Smith V, Peng A, Chiou WL. Simple and rapid high pressure liquid chromatographic simultaneous determination of aspirin, salicylic acid and salicyluric acid in plasma. J Pharmaceut Sci 1978; 67: 710–712.
14. Levy G, Procknall JA, Olufs R, Pachman LM. Relationship between saliva salicylate concentration and free or total salicylate concentration in serum of children with juvenile rheumatoid arthritis. Clin Pharmacol Ther 1980; 27: 619–627.
15. Leonards JR. Presence of acetylsalicylic acid in plasma following oral ingestion of aspirin. Proc Soc Exp Biol Med 1962; 110: 304–307.
16. Rowland M, Riegelman S, Harris PA, Sholkoff SD. Absorption kinetics of aspirin in man following oral administration of an aqueous solution. J Pharmaceut Sci 1972; 61: 379–385.
17. Volans GN. Effects of food and exercise on the absorption of effervescent aspirin. Br J Clin Pharmacol 1974; 1: 137–141.
18. Leonards JR. The influence of solubility on the rate of gastrointestinal absorption of aspirin. Clin Pharmacol Ther 1963; 4: 476–479.
19. Martin BK. The formulation of aspirin. In: Bean J (ed.) Advances in Pharmaceutical Sciences, Vol 3. London: Academic Press, 1971: 107–171.
20. Castleden CM, Volans CM, Raymond K. The effect of ageing on drug absorption from the gut. Age Ageing 1977; 6: 138–143.
21. Salem SA, Stevenson IH. Absorption kinetics of aspirin and quinine in elderly subjects. Br J Clin Pharmacol 1977; 4: 397P.
22. Cuny G, Royer RT, Mur JM *et al.* Pharmacokinetics of salicylates in the elderly. Gerontology 1979; 25: 41–55.
23. Morgan AM, Truitt EB Jr. Evaluation of acetylsalicylic acid esterase in aspirin metabolism. J Pharmaceut Sci 1965; 54: 1640–1666.
24. Ross Lee LM, Elms MJ, Cham BE *et al.* Plasma levels of aspirin following effervescent and enteric coated tablets, and their effect on platelet function. Eur J Clin Pharmacol 1982; 23: 545–551.
25. Editorial. Poisoning with enteric-coated aspirin. Lancet 1981; 2: 130.

26. Clissold SP. Aspirin and related derivatives of salicylic acid. Drugs 1986; 32 (suppl 4): 8–26.
27. Graham GG, Champion GD, Day RO, Paull PD. Patterns of plasma concentrations and urinary excretion of salicylate in rheumatoid arthritis. Clin Pharmacol Ther 1977; 22: 410–420.
28. Insel PA. Analgesic-antipyretic and antiinflammatory agents: the salicylates. In: Hardman JG, Limbird LE (eds) Goodman and Gilman's The Pharmacological Basis of Therapeutics, 9th edn. New York: Mosby, 1996: 617–657.
29. Abdallah HY, Mayersohn M, Conrad KA. The influence of age on salicylate pharmacokinetics in humans. J Clin Pharmacol 1991; 31: 380–387.
30. Roberts R, Rumble RH, Wanwimolruk S, Thomas D, Brooks PM. Pharmacokinetics of aspirin and salicylate in elderly subjects and in patients with alcoholic liver disease. Eur J Clin Pharmacol 1983; 25: 253–261.
31. Wanwimolruk S, Birkett DJ, Brooks PM. Protein binding of some nonsteroidal anti-inflammatory drugs in rheumatoid arthritis. Clin Pharmacokinetics 1982; 7: 85–92.
32. Borga O, Odar-Cederlof I, Ringberger V, Norlin A. Protein binding of salicylate in uraemic and normal plasma. Clin Pharmacol Ther 1976; 20: 464–475.
33. Levy G, Baliah T, Procknal JA. Effect of renal transplantation on protein binding of drugs in serum of donor and recipient. Clin Pharmacol Ther 1976; 20: 512–516.
34. Levy G, Procknall JA, Garrettson LK. Distribution of salicylate between neonatal and maternal serum at diffusion equilibrium. Clin Pharmacol Ther 1975; 18: 210–214.
35. Hawkins D, Pinckard RN, Farr RS. Acetylation of human serum albumin by acetylsalicylic acid. Science 1968; 160: 780–781.
36. Chen CN, Coleman DL, Andrade JD, Temple AR. Pharmacokinetic model for salicylate in cerebrospinal fluid, blood, organs and tissues. J Pharmaceut Sci 1978; 67: 38–45.
37. Lorenzo AV, Spector R. Transport of salicylic acid by the choroid plexus in vitro. J Pharmacol Exp Ther 1973; 184: 465–471.
38. Rosenthal RK, Bayles TB, Fremont Smith K. Simultaneous salicylate concentrations in synovial fluid and plasma in rheumatoid arthritis. Arthritis Rheum 1964; 7: 103–109.
39. Trnavska Z, Trnavska K. Characterisation of salicylate binding to synovial fluid and plasma protein in patients with rheumatoid arthritis. Eur J Clin Pharmacol 1980; 18: 403–406.
40. Soren A. Kinetics of salicylates in blood and joint fluid. Eur J Clin Pharmacol 1979; 16: 279–285.
41. Findlay JW, DeAngelis RL, Kearney MF, Welch RM, Findlay JM. Analgesic drugs in breast milk and plasma. Clin Pharmacol Ther 1981; 29: 625–633.
42. Harris PA, Riegelman S. Acetylsalicylic acid hydrolysis in human blood and plasma. J Pharmaceut Sci 1967; 56: 713–716.
43. Rainsford KD. Aspirin and the Salicylates. London: Butterworths, 1984.

44. Williams FM, Wynne H, Woodhouse KW, Rawlins MD. Plasma aspirin esterase: the influence of old age and frailty. Age Ageing 1989; 18: 39–42.
45. O'Mahoney MS, George G, Westlake H, Woodhouse KW. Plasma aspirin esterase activity in elderly patients undergoing elective hip replacement and with fractured neck of femur. Age Ageing 1994; 23: 338–341.
46. Williams FM, Nicholson EN, Woodhouse KW, Adjepon-Yamoah KK, Rawlins MD. Activity of esterases in plasma from Ghanian and British subjects. Eur J Clin Pharmacol 1986; 31: 485–489.
47. Markiewicz A, Semenowiczk A. Time dependent changes in the pharmacokinetics of aspirin. Int J Clin Pharmacol Biopharm 1979; 17: 409–411.
48. Dawson I, Pryse-Davies J. The distribution of certain enzyme systems in the normal human gastrointestinal tract. Gastroenterology 1963; 44: 745–760.
49. Rowland M, Riegelman S. Pharmacokinetics of acetylsalicylic acid and salicylic acid after intravenous administration in man. J Pharmaceut Sci 1968; 57: 1313–1319.
50. Hutt AJ, Caldwell J, Smith RL. The metabolism of [carboxyl-14C]aspirin in man. Xenobiotica 1982; 12: 601–610.
51. Wilson JT, Howell RL, Holladay MW *et al.* Gentisuric acid: Metabolic formation in animals and identification as a metabolite in man. Clin Pharmacol Ther 1978:23; 635–643.
52. Ho PC, Triggs EJ, Bourne DW, Heazlewood VJ. The effects of age and sex on the disposition of acetylsalicylic acid and its metabolites. Br J Clin Pharmacol 1985; 19: 675–684.
53. Hutt AJ, Caldwell J, Smith RL. The metabolism of aspirin in man: a population study. Xenobiotica 1986: 16; 239–249.
54. Levy G, Tsuchiya T. Salicylate accumulation kinetics in man. N Engl J Med 1972; 287: 430–432.
55. Patel DK, Hesse A, Ogunbona A, Notarianni LJ, Bennett PN. Metabolism of aspirin after therapeutic and toxic doses. Hum Exp Toxicol 1990; 9: 131–136.
56. Paulus HE, Siegel M, Morgan E, Okun R, Calabro JJ. Variations in serum concentrations and half-life of salicylate in patients with rheumatoid arthritis. Arthritis Rheum 1971; 14: 527–532.
57. Day RO, Shen DD, Azarnoff DL. Induction of salicyluric acid formation in rheumatoid arthritis patients treated with salicylates. Clin Pharmacokinetics 1983; 8: 263–271.
58. Day RO, Furst DE, Dromgoole SH, Paulus HE. Changes in salicylate serum concentration and metabolism during chronic dosing in normal volunteers. Biopharm Drug Dispos 1988; 9: 273–283.
59. Owen SG, Roberts MS, Friesen WT, Francis HW. Salicylate pharmacokinetics in patients with rheumatoid arthritis. Br J Clin Pharmacol 1989; 28: 449–461.
60. Tsuchiya T, Levy G. Biotransformation of salicylic acid and its acyl and glucuronides in man. J Pharmaceut Sci 1972; 61: 800–801.
61. Hansten PD, Hayton WL. Effect of antacid and ascorbic acid on serum salicylate concentrations. J Clin Pharmacol 1980; 20: 326–331

62. Shastri RA. Effect of antacids on salicylate kinetics. Int J Clin Pharmacol Ther Toxicol 1985; 23: 480–484.
63. Levy G, Leonards JR. Urine pH and salicylate therapy. JAMA 1971; 217: 81.
64. International Stroke Trial Collaborative Group. The International Stroke Trial (IST): a randomised trial of aspirin, subcutaneous heparin, both, or neither among 19435 patients with acute ischaemic stroke. Lancet 1997; 349: 1569–1581.
65. Committee on Safety of Medicines/Medicines Control Agency. Paracetamol and aspirin. Curr Prob Pharmacovigilance 1997; 23: 9.
66. Orton D, Treharne Jones R, Kapsi T, Richardson R. Plasma salicylate levels after soluble and effervescent aspirin. Br J Clin Pharmacol 1979;7: 410–412.
67. Gatti G, Barzaghi N, Attardo Parrinello G, Vitiello B, Perucca E. Pharmacokinetics of salicylic acid following administration of aspirin tablets and three different forms of soluble aspirin in normal subjects. Int J Clin Pharmacol Res 1989; 9: 385–389.
68. Kelly CA. Determination of the decomposition of aspirin. J Pharmaceut Sci 1970; 59: 1053–1078.
69. Thiessen JJ. Pharmacokinetics of salicylates. In Barnett HJM, Hirsh J, Mustard JF (eds). Acetylsalicylic Acid: New Uses for an Old Drug. New York: Raven Press, 1982: 49–61.
70. Brooks PM, Roberts MS, Patel P. Pharmacokinetics of sustained release aspirin. Br J Pharmacol 1978; 5: 337–339.
71. Anttila M, Kahela P, Uotila P. The absorption of acetylsalicylic acid from enteric-coated formulation and the inhibition of thromboxane formation. Int J Pharmacol Ther Toxicol 1988; 26: 88–92.
72. Montgomery PR, Sitar DS. Acetylsalicylic acid metabolites in blood and urine after plain and enteric-coated tablets. Biopharm Drug Dispos 1986; 7: 21–25.
73. Leonards JR, Levy G. Absorption and metabolism of aspirin administered in enteric coated tablets. JAMA 1965; 193: 93–98.
74. Blythe RH, Grass GM, MacDonnell DR. The formulation and evaluation of enteric coated aspirin tablets. Am J Pharmacol 1959; 131: 206–216.
75. Anslow JA, Greene DS, Hooper JW, Wagner GS. A new enteric coated preparation of aspirin. Curr Ther Res 1984; 36: 811–818.
76. Hofteizer JV, Silvoso GR, Burks M, Ivey KJ. Comparison of the effects of regular and enteric coated aspirin on gastroduodenal mucosa of man. Lancet 1980; 2: 609–612.
77. Lanza FL, Royer GL Jr, Nelson RS. Endoscopic evaluation of the effects of aspirin, buffered aspirin, and enteric coated aspirin on gastric and duodenal mucosa. N Engl J Med 1980; 303:136–138.
78. Hawthorne AB, Mahida YR, Cole AT, Hawkey CJ. Aspirin-induced gastric mucosal damage: prevention by enteric-coating and relation to prostaglandin synthesis. Br J Clin Pharmacol 1991; 32: 77–83.

79. Savon JJ, Allen ML, DiMarino AJ, Hermann GA, Krum RP. Gastrointestinal blood loss with low dose (325mg) plain and enteric-coated aspirin administration. Am J Gastroenterol 1995; 90: 581–585.
80. Silagy CA, McNeil JJ, Donnan GAet al. Adverse effects of low-dose aspirin in a healthy elderly population. Clin Pharmacol Ther 1993; 54: 84–89.
81. Kelly JP, Kaufman DW, Jurgelon JN *et al.* Risk of aspirin-associated major upper gastrointestinal bleeding with enteric or buffered product. Lancet 1996; 348: 1314–1316.
82. Guslandi M. Gastric toxicity of anti-platelet therapy with low-dose aspirin. Drugs 1997; 53: 1–5.
83. Petroski D. Gastric safety and enteric-coated aspirin. Lancet 1997; 349: 430–431.
84. Levy G, Hayes BA. Physicochemical basis of the buffered acetylsalicylic acid controversy. N Engl J Med 1960; 262:1053.
85. Graham DY, Smith JL. Aspirin and the stomach. Ann Intern Med 1986; 104: 390–398.
86. Javaid KA, Cadwallader DE. Dissolution of aspirin from tablets containing various buffering agents. J Pharmaceut Sci 1972; 61: 1370–1373.
87. Mason WD, Winer N. Kinetics of aspirin, salicylic acid and salicyluric acid following oral administration of aspirin as a tablet and two buffered solutions. J Pharmaceut Sci 1981; 70: 262–265.
88. Mason WD, Winer N. Influence of food on aspirin absorption from tablets and buffered solutions. J Pharmaceut Sci 1983; 72: 819–821.
89. Silvoso GR, Ivey KJ, Butt JH *et al.* Incidence of gastric lesions in patients with rheumatic disease on chronic aspirin therapy. Ann Intern Med 1979; 91: 517–520.
90. Fries JF, Ramey DR, Singh G *et al.* A reevaluation of aspirin therapy in rheumatoid arthritis. Arch Intern Med 1993; 153: 2465–2471.
91. Brandslund I, Rask H, Klitgaard NA. Gastrointestinal blood loss caused by controlled-release and conventional acetylsalicylic acid tablets. Scand J Rheumatol 1979; 8: 209–213.
92. Dybdahl JH, Daae LN, Larsen Set al. Acetylsalicylic acid-induced gastrointestinal bleeding determined by a 51Cr method on a day-to-day basis. Scand J Gastroenterol 1980; 15: 887–895.
93. Roberts MS, Joyce RM, McLeod LJ, Vial JH, Seville PR. Slow-release aspirin and prostaglandin inhibition. Lancet 1986; i: 1153–1154.
94. Vial JH, McLeod LJ, Roberts MS, Seville PR. Selective inhibition of platelet cyclooxygenase with controlled release, low-dose aspirin. Aust NZ J Med 1990; 20: 652–656.
95. Clarke RJ, Mayo G, Price P, Fitzgerald GA. Suppression of thromboxane A2 but not of systemic prostacyclin by controlled release aspirin. N Engl J Med 1991; 325: 1137–1141.
96. Gibaldi M, Grundhofer B. Bioavailability of aspirin from commercial suppositories. J Pharmaceut Sci 1975; 64: 1064–1066.
97. Levy G. Clincal pharmacokinetics of aspirin. Pediatrics 1978; 62 (suppl): 867–872.

98. Keimowitz RM, Pulvermacher G, Mayo G, Fitzgerald DJ. Transdermal modification of platelet function: a dermal aspirin preparation selectively inhibits platelet cyclo-oxygenase and preserves prostacyclin synthesis. Circulation 1993; 88: 556–561.
99. De Benedittis G, Lorenzetti A. Topical aspirin/diethyl ether mixture versus indomethacin and diclofenac/diethyl ether mixtures for acute herpetic neuralgia and postherpetic neuralgia: a double-blind crossover placebo-controlled study. Pain 1996; 65: 45–51.
100. Tharion G, Bhattacharji S. Aspirin in chloroform as an effective adjuvant in the management of chronic neurogenic pain. Arch Phys Med Rehab 1997; 78: 437–439.
101. Yosipovitch G, Ademola J, Lui P, Amin S, Maibach HI. Topically applied aspirin rapidly decreases histamine-induced itch. Acta Dermatol Venereol 1997; 77: 46–48.
102. Anonymous. Which prophylactic aspirin? Drugs Therapeut Bull 1997; 35: 7–8.
103. Levy G. Pharmacokinetics of salicylate elimination in man. J Pharmaceut Sci 1965; 54: 959–967.
104. Smith PK, Gleason HL, Stoll CG, Ogorzaleks S. Studies on the pharmacology of salicylates. J Pharmacol 1946; 87: 237–255.

Mode of action and pharmacological effects

Introduction

Successful therapeutic use of a drug is generally facilitated by a thorough knowledge of its mode of action. In the case of aspirin, it managed to maintain for 70 years its pre-eminent position as the anti-inflammatory, antirheumatic and analgesic agent of choice, despite little understanding of the way in which it worked [1]. The discovery of the inhibitory role of aspirin on the synthesis and release of prostaglandins [2] led to the development of numerous other safer and possibly more efficacious non-steroidal anti-inflammatory drugs and encouraged interest in its potential therapeutic use in prophylaxis and treatment of vascular disease, pregnancy and cancer. It remains a widely used household remedy and a powerful pharmacological tool in the investigation of mechanisms of pain and inflammation and arterial thrombosis.

Mode of action

Most of the important pharmacological effects of aspirin *in vivo* are thought to be based on the inhibition of prostaglandin synthesis, due to the selective acetylation of the hydroxyl group of a single serine residue at position 530 within the polypeptide chain of the enzyme cyclo-oxygenase (COX), also known as prostaglandin H2 synthase [3, 4]. This enzyme catalyses the first stage of prostaglandin synthesis, the oxygenation of arachidonic acid to prostaglandin G2 and then its subsequent reduction to prostaglandin H2, which is then further metabolized to other prostaglandins, prostacyclin and thromboxane A2, which collectively are called eicosanoids [5] (Figure 3.1).

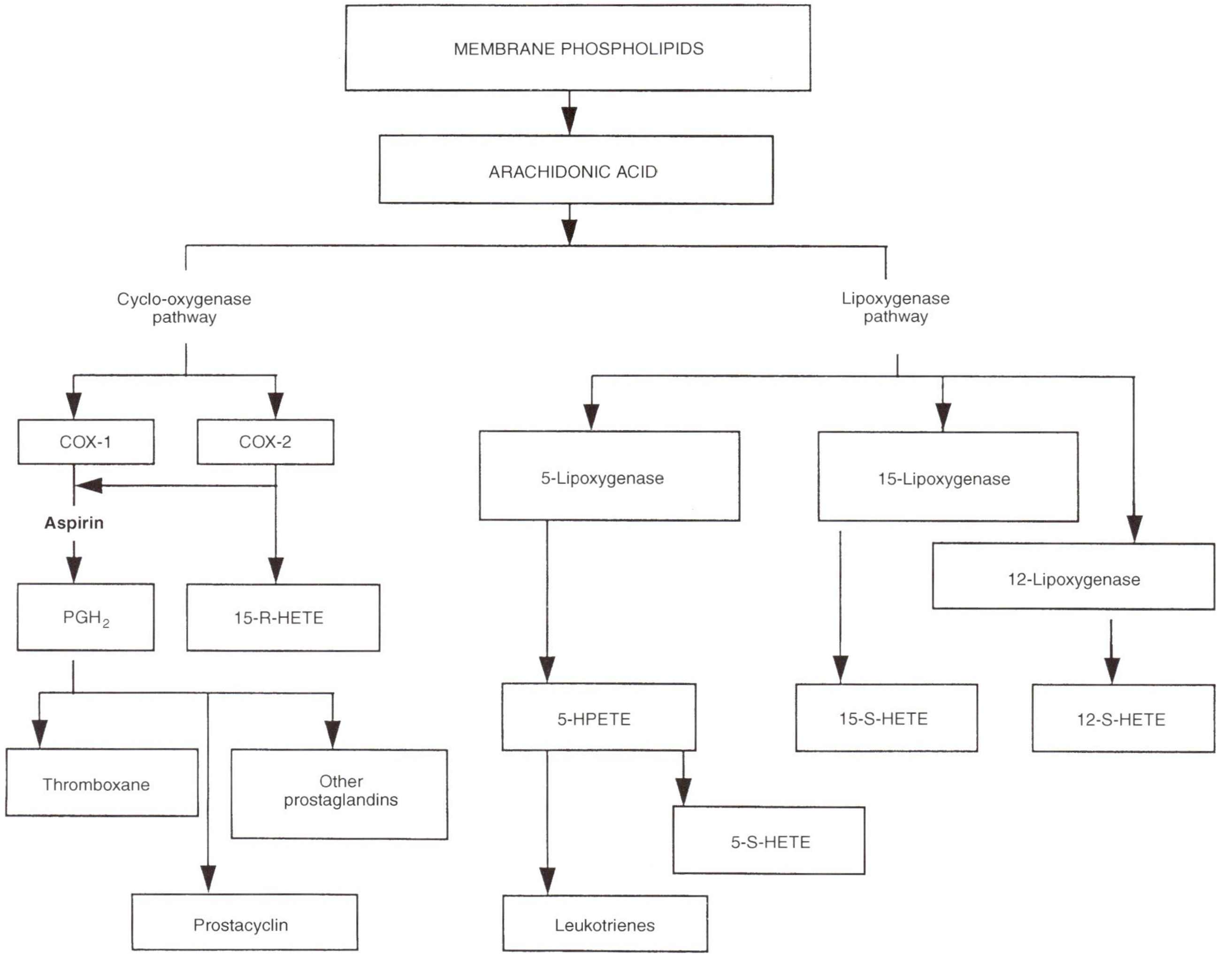

Fig. 3.1
The metabolism of arachidonic acid, via the cyclo-oxygenase and lipoxygenase pathways, and the site of action of aspirin.

Acetylation of COX is irreversible and new enzyme must be synthesized before prostaglandin production can be resumed.

There are at least two distinct isoforms of COX, known as COX-1 and COX-2, which share 60% amino acid sequence identity. A third form, COX-3, is also postulated [6]. COX-1 is present in various tissues in significant amounts and appears to have a variety of homoeostatic functions including gastric mucosal cytoprotection, control of renal perfusion and antithrombogenic effects. COX-2 is present only in low concentrations in normal tissue, but is induced by proinflammatory stimuli such as release of cytokines and endotoxins and produces inflammatory prostaglandins. Aspirin is a strong inhibitor of COX-1, but has less potent effects against COX-2. It is the inhibition of COX-1 which is responsible for the beneficial antiplatelet action and for many of the adverse effects of aspirin, particularly its gastrointestinal toxicity. Aspirin-acetylated COX-2 converts arachidonic acid to 15-*R*-hydroxyeicosatetraenoic acid (15-*R*-HETE) instead of prostaglandin G2, an action which is thought may be relevant to its anti-inflammatory and anti-tumour properties [7].

Recent studies using X-ray crystallography to demonstrate the molecular structure of COX-1 inactivated by the aspirin analogue, 2-bromoacetoxybenzoic acid, have shown that aspirin acts by blocking or partially blocking the substrate binding site [8]. Thus the acetyl group introduced by aspirin to the side chain of the serine-530 residue extends across the channel leading to the active binding site, preventing arachidonic acid from binding productively or causing abnormal binding and production of 15-*R*-HETE [9].

Other non-prostaglandin mechanisms of action of aspirin have also long been suggested, but their relative importance is uncertain. A recent review was largely dismissive, referring to 'outposts of disaffection clinging to ambiguous modes of action . . . despite the satisfaction afforded by a simple and unifying theory' [10]. Certainly many of the reported effects have only been observed *in vitro* rather than *in vivo* [11–13].

ANTI-INFLAMMATORY EFFECTS

Until well into the second half of the 20th century, it was the potent anti-inflammatory properties of aspirin which primarily justified its pre-eminent position in medical treatment. High concentrations of prostaglandins are found at sites of inflammation, such as in experimentally induced inflammatory arthritis in animals [15] and in the joints of rheumatoid arthritis patients [16], and the relatively low pH extracellular environment of the synovial fluid will favour the accumulation of acidic aspirin in joints due to the phenomenon of 'ion trapping' [17]. Thus inhibition of prostaglandin synthesis is an attractive mechanism to explain its potent anti-inflammatory effects [2,14]. However, aspirin will not block the effects of prostaglandins which have already been released and the inhibitory effect on prostaglandin synthesis is soon reversed as new COX is synthesized.

There is also powerful evidence that the effects on prostaglandin production may only partly explain the anti-inflammatory actions of aspirin. For example, the dose which is required to suppress inflammation significantly exceeds the dose required to inhibit prostaglandin synthesis, at least in plasma [12]. Furthermore, sodium salicylate, which is much less potent than aspirin in inhibiting prostaglandin synthesis *in vitro* [18], is nevertheless as effective an anti-inflammatory agent in laboratory models [19] and in patients with rheumatoid arthritis [20,21] Indeed, overall it has been concluded that the acetyl group of aspirin contributes little to its anti-inflammatory effect [22–24].

Possible non-prostaglandin mediated mechanisms which have been suggested as contributing to aspirin's anti-inflammatory effects include the following.

- The process of binding of salicylate to plasma proteins displaces tryptophan and related peptides, which may themselves have anti-inflammatory properties [25].
- Salicylate inhibits the migration of polymorphonuclear leucocytes and monocytes into inflamed tissues and interferes with their function, including generation of prostaglandins, leukotrienes and other substances important in the inflammatory cycle [26,27].
- Salicylate has the ability to buffer free radicals generated as part of the inflammatory response [28,29].
- Aspirin-like drugs inhibit various lipoxygenase enzymes [30] and may prevent metabolism of arachidonic acid along the alternative pathway to the proinflammatory lipoxins. These biologically active eicosanoids participate in the inflammatory response and may be involved with consequent leukotriene production [31].
- Aspirin, but not salicylate, inhibits another component of the inflammatory response, the complement activation pathway [32].
- Stimulation of the pituitary–adrenal system may occur with high doses of aspirin, causing production of anti-inflammatory corticosteroids [33].
- Salicylate may interfere with cell membrane-related processes, including the activity of phospholipase C in macrophages [34] and the release of lysosomal enzymes, by stabilizing the lysosomal membrane [29,35]. It may also bind to key regulatory G proteins in the cell membrane, interfering with signal transduction in a variety of cells, including the neutrophil [36].
- Salicylate may inhibit the increased extracellular proteolytic activity often associated with inflammatory processes [37].

- Salicylates influence cartilage metabolism, by interfering with the synthesis of proteoglycans [38].

The relative importance, if any, of each of these suggested mechanisms is uncertain.

ANALGESIC EFFECTS

Aspirin appears to produce analgesia by both peripheral and central mechanisms. Its anti-inflammatory action may also play a role [39,40]. The most important action appears to be peripheral, occurring at the site of the painful or inflammatory stimulus [41,42].

In response to tissue damage, nociceptors (pain receptors) in peripheral nerves are normally sensitized to the hyperalgesic effects of prostaglandins, which also cause them to be more sensitive to the pain-producing effects of other substances such as serotonin, adenosine, histamine and bradykinin. This effect will be blocked by the inhibition of synthesis and release of prostaglandins by the acetylating action of aspirin. Thus aspirin would be expected to be most effective against the dull, throbbing pain of inflammation and less effective against the sharp, stabbing pain caused by direct stimulation of sensory nerves. Certainly in animal experiments aspirin raises the pain threshold when pressure is applied to a swollen inflamed foot, but not when applied to a normal foot [43]. Clinically, it is particularly active in treating pain associated with inflammation but is ineffective, for example, in pain associated with nerve compression [44,45].

Salicylate itself, which has no inhibitory effect on COX when given in standard doses, has much less analgesic effect than aspirin [46]. Thus in pharmacokinetic/pharmacodynamic models, the persistence of the unhydrolysed ester in blood corresponds approximately to the duration of analgesia, with salicylic acid being non-contributory. Peak analgesic effect occurs 20–30 minutes after time to peak plasma aspirin concentration [47]. The higher the aspirin esterase activity in plasma, the lower the amount of analgesia produced [48]. The effect of frailty and injury on aspirin esterase activity [49] may therefore be of relevance when aspirin is used for pain relief in elderly patients.

Non-prostaglandin mediated mechanisms may also play some role in aspirin's analgesic effect as the non-acetylated salicylates which do not inhibit prostaglandin production nevertheless do possess some pain-relieving activity [21]. There is certainly evidence that some of the analgesic properties of aspirin may be due to a direct action within the central nervous system. In animals, non-steroidal anti-inflammatory drugs inhibit spinal cyclo-oxygenase and block hyperalgesia [50] and an effect on the hypothalamus has also been postulated [51,52]. In general, however, the absence of central nervous system side effects associated with analgesic doses of aspirin suggests that the contribution of any central activity must be small. Furthermore, even regular high doses of aspirin do not induce tolerance or addiction.

ANTIPYRETIC EFFECTS

In therapeutic doses, aspirin has no effect on normal body temperature. Its antipyretic effects in febrile patients, however, are well known [53]. Fever associated with infection is caused by release of cytokines and other inflammatory mediators, mainly from white blood cells. These act on the central thermostat in the hypothalamus, causing production of prostaglandin E2 which 'resets' the body temperature to a higher level [54]. Inhibition of prostaglandin synthesis is thought to be primarily responsible for aspirin's antipyretic effect [55,56], though heat dissipation will also be increased as a result of vasodilatation and increased peripheral blood flow. Sodium salicylate, which has little antiprostaglandin effect, has about half the antipyretic potency of aspirin, on a molar basis [57].

Moderate doses of aspirin that effectively lower body temperature will also increase oxygen consumption and metabolic rate and consequently, when taken in toxic doses, salicylates themselves can have a pyretic effect. High therapeutic doses of aspirin may also impair the acclimatization to work and exercise in hot temperatures [58,59].

ANTITHROMBOTIC EFFECTS

Aspirin prolongs the bleeding time and decreases platelet aggregation, a property which is the basis for aspirin's widespread use for antithrombotic prophylaxis and treatment [60,61]. Even before the discovery of aspirin's antiplatelet effect, salicylic acid in high dosage was known to inhibit the synthesis of prothrombin and other vitamin K-dependent coagulation factors [62,63], though this is rarely of significance except in cases of overdose or in patients with severe hepatic disease. Aspirin also possesses some fibrinolytic properties, possibly due to reduction of plasminogen activator activity, but this is probably not of any clinical significance [64,65].

The important antiplatelet effect of aspirin is primarily related to inhibition of platelet COX-1, resulting in reduced conversion of arachidonic acid to prostaglandins G2 and H2 and so to decreased synthesis of thromboxane A2. It is thromboxane which is normally activated by stimuli such as thrombin, collagen and adenosine diphosphate and which induces irreversible platelet aggregation.

The effect of aspirin on COX-1 is rapid and predominantly occurs as platelets pass through the portal circulation, before the drug reaches the systemic circulation [66]. Salicylate does not inhibit platelet function [67] and a relationship has been demonstrated between activity of plasma aspirin esterase and the effect of aspirin on bleeding time [68]. As platelets are anucleate and unable to synthesize new proteins, the antiplatelet effect is permanent and cannot be reversed during the platelet's 8–10-day lifespan. Recovery in COX-1 activity is dependent on new platelets being produced from megakaryocyte precursors. A single dose of aspirin will therefore affect platelet function for several days and the inhibitory effect of repeated daily doses will be cumulative. Furthermore, the response is dose dependent within the range from 5 to 100 mg, with the higher dose almost completely suppressing thromboxane A2 production in healthy subjects [69] and in atherosclerotic patients [70]. Typically, aspirin prolongs bleeding time by about two minutes, but some 14% of subjects appear to be 'hyper-responders' and may be at greater risk of bleeding complications [71].

A possible effect of low-dose aspirin on platelet-activating factor (PAF), the proaggregatory agent which is formed from the lysophospholipids released along with arachidonic acid as a result of membrane phospholipid breakdown, has also been hypothesized but not substantiated [72].

Higher dose aspirin (> 300 mg) acetylates fibrinogen and enhances fibrinolysis [73], facilitating plasminogen activation and attenuating platelet aggregation [74]. Thrombin formation in clotting blood is also reduced [75] and very large doses may cause significant hypoprothrombinaemia. Even a single 1000 mg oral dose of aspirin may result in a statistically significant reduction in prothrombin time [76].

Whilst these effects might prevent or limit thrombus formation and favour haemorrhage, there may also be prothrombotic effects. Acetylation of endothelial COX-1 reduces synthesis of prostacyclin, a potent vasodilator and inhibitor of platelet aggregation [77] and acetylation of macrophage COX-2 may result in changes in leukotrienes and lipoxin which may diminish resistance to thrombosis [78]. However, as both endothelium and macrophages can regenerate COX activity, the predominant effect of aspirin will generally be on thromboxane A2 synthesis, with only brief and intermittent inhibition of prostacyclin. It is generally believed that it is the balance between thromboxane and prostacyclin which normally maintains vascular integrity [79]. The lower the aspirin dose, the greater the differential effect, as small amounts of aspirin in the presystemic circulation will irreversibly inhibit platelets but will not be adequate to have a significant effect on endothelium on the systemic side of the circulation. There is, however, no dose which is absolutely selective.

OTHER HAEMATOLOGICAL EFFECTS

As well as its antithrombotic action, aspirin also has some indirect haematological effects. Thus chronic gastrointestinal blood loss associated with prolonged daily use of aspirin, although not usually of clinical significance, may lead to depletion of iron stores and eventually to iron deficiency anaemia. This will depend on the rate of blood loss and the initial size of the iron load and there is much individual variation. One aspirin tablet (325 mg) will typically cause blood loss amounting to 1–10 ml per day, or 0.5–5 mg iron loss per day. Adults without chronic blood loss store 300–1000 mg of iron and so this can become exhausted within months [80]. However, iron depletion has itself been suggested as an important second mode of action of aspirin used long term, rather than an undesirable side effect. Loss of stored iron is claimed to be protective against heart disease [81] and possibly cancer [82] and the cumulative protective effect of aspirin would be consistent with benefit arising from the progressive lessening of iron stores over time.

GASTROINTESTINAL EFFECTS

Prostaglandins, particularly of the E series, play an important role in the maintenance of normal gastrointestinal physiology, potentially protecting the gastric mucosa by inhibiting acid secretion, raising bicarbonate output and mucus secretion, maintaining mucosal blood flow and controlling mucosal permeability to hydrogen ions [83,84]. By inhibiting the activity of gastric mucosal COX-1, aspirin interferes with the synthesis of these prostaglandins. This removes from the gastric luminal surface their protective effect and the mucosa is put in jeopardy by a combination of increased acid production, decreased mucus production, decreased gastric blood flow and back diffusion of hydrogen ions into the gastric lumen [85,86]. There is consequent loss of mucosal integrity, which manifests as gastric erosions, ulceration and bleeding. Further support for the central role of prostaglandin inhibition in producing aspirin-induced injury is given by the finding that it can be prevented by co-administration of prostaglandin E2 [87,88] or the prostaglandin analogue misoprostol [89], which will also help to heal aspirin-induced gastroduodenal injury in patients with rheumatoid arthritis [90].

Other mechanisms which may play a small role include a direct irritant effect causing aggregation and sloughing of the mucosal layer and destruction of the underlying cells [91] and interference with platelet function causing focal haemorrhage.

Whilst animal studies have shown that high-dose systemic aspirin reaching the stomach through the circulation can also cause damage [88], it is generally accepted that it is the topical effect of aspirin which is most important [92]. By minimizing contact time between the drug and the gastric mucosa, buffered, enteric-coated or modified-release preparations of aspirin may offer some small advantages in terms of gastrointestinal tolerance, especially if long-term high-dose therapy is necessary [93]. Recent animal studies have also suggested that aspirin-induced gastrointestinal injury can be reduced, while enhancing anti-inflammatory and antipyretic effects, by preassociating aspirin with phospholipids to form a zwitterionic compound [94]. This is assumed to be secondary to its increased lipid solubility and permeability. The potential clinical benefits of such phospholipid enhancement have yet to be investigated. Gastric irritation is not the only mechanism by which aspirin may cause gastrointestinal symptoms. When plasma salicylate levels are above 250 mg/l (1.8 mmol/l), the chemoreceptor trigger zone in the medulla is directly stimulated and nausea and vomiting may ensue [95].

RENAL EFFECTS

Aspirin has much less effect on the kidney than other non-steroidal anti-inflammatory drugs, but nevertheless the blockade of intrarenal prostaglandins may lead to decline in renal blood flow and fall in glomerular filtration rate with water, sodium and potassium retention [96,97]. This may be critical in those with already compromised function and patients with systemic lupus erythematosus or Still's disease are said to be at particular risk [97, 98]. Certainly, even a single oral dose of 1000 mg aspirin given to normal human volunteers significantly increases both beta 2-microglobulin and albumin excretion rates in urine within two hours of dosage [76]. Regular use of therapeutic doses can produce a transient increase in urinary excretion of renal tubular epithelial cells, increases in blood urea and significant proteinuria [44]. However, intermittent use of anti-inflammatory doses of aspirin and long-term prophylactic use of low doses is probably rarely, if ever, associated with significant nephrotoxicity [99–101].

CARDIOVASCULAR AND RESPIRATORY EFFECTS

Therapeutic doses of aspirin generally have no direct cardiovascular effects. Large doses, especially when taken in toxic amounts, may have a direct effect on smooth muscle in peripheral vessels, resulting in vasodilatation. They may also lead to an increase in circulating plasma volume and subsequent pulmonary oedema. This used to be seen in older patients with compromised cardiac function who were taking high doses of aspirin regularly but it is now rare. Central vasomotor paralysis associated with aspirin toxicity has also been described, leading to cardiovascular collapse [95].

Effects on respiration may be seen with anti-inflammatory and toxic doses of aspirin (Figure 3.2). The uncoupling of oxidative phosphorylation increases consumption of oxygen and production of carbon dioxide and this stimulates the depth and to a lesser extent the rate of ventilation [102]. High doses of aspirin resulting in plasma salicylate concentrations greater than 350 mg/ml (2.57 mmol/l) also directly stimulate the respiratory centre, resulting in an increase in both depth and rate of ventilation. Consequently, respiratory alkalosis will develop. Marked hyperpnoea occurs with plasma salicylate levels approaching 500 mg/ml (3.66 mmol/l) [95]. Yet higher doses of aspirin or prolonged exposure to toxic salicylate levels have a central respiratory depressant effect and, as enhanced carbon dioxide production persists, this may result in respiratory acidosis.

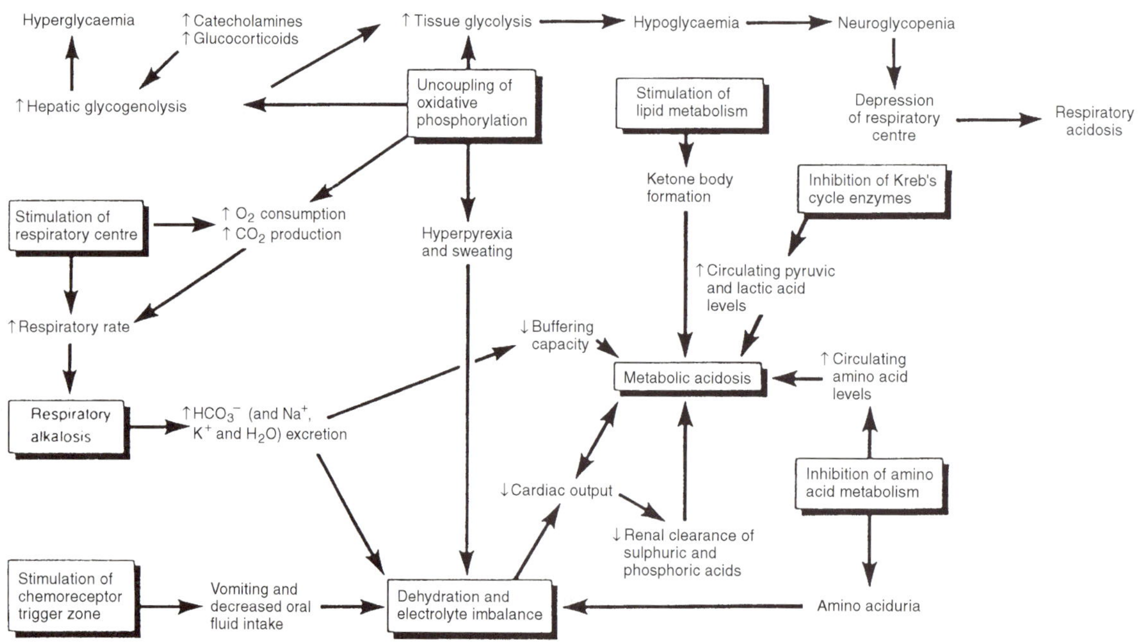

Fig. 3.2
Metabolic effects of toxic doses of aspirin: the pathophysiology of salicylate poisoning (from reference [121] with permission).

METABOLIC AND ENDOCRINE EFFECTS

The metabolic and endocrine effects of aspirin are complex but are generally not important unless high doses are taken. The acute toxicity of salicylates is largely a result of metabolic disturbance. The effects include changes in acid–base and water and electrolyte balance [95], inhibition of various enzyme systems including those responsible for oxidative phosphorylation, the pyridine nucleotide-dependent dehydrogenases and certain oxidases [103,104] and interference with amino acid, lipid and carbohydrate metabolism [102]. Salicylate-induced inhibition of xanthine oxidase activity may complement its uricosuric effects and contribute to its clinical effects in gout. The endocrine effects may involve the hypothalamic–pituitary axis, the adrenal and thyroid glands [105]. All of these effects will occur simultaneously. The net effect is therefore not easily predictable and is the result of a mixture of superimposed actions on several different systems.

The uncoupling of oxidative phosphorylation by salicylates may occur at anti-inflammatory dose levels and at one stage was considered partly responsible for the therapeutic effects [106]. It results in failure to produce high energy phosphates, with inhibition of a number of adenosine triphosphate (ATP)-dependent reactions, increase in oxygen uptake and carbon dioxide production and depletion of hepatic stores of glycogen. In toxic doses, the energy normally used to produce ATP will be dissipated as heat, leading to pyrexia and sweating (Figure 3.2).

The effects of aspirin on blood glucose levels may lead to hypo- or hyperglycaemia. Hypoglycaemia may develop through a variety of mechanisms, including stimulation of insulin secretion [107], uncoupling of oxidative phosphorylation [104], suppression of free fatty acid release from adipose tissue [108] and a possible effect on glycaemic control centres in brain and spinal cord [109]. Significant effects have generally only been reported in those taking high doses or in patients with renal failure [110,111]. However, therapeutic doses of salicylates can lower blood glucose concentrations in both diabetic and non-diabetic subjects and well into the second half of the century aspirin was even being promoted as a possible alternative to insulin therapy [112]. Conversely, very high therapeutic and toxic doses of aspirin may provoke hyperglycaemia. This arises due to increased glucose-6-phosphatase activity, release of adrenalin, which depletes liver and muscle stores of glycogen, and increase in glucocorticoid secretion [102].

Stimulation of the hypothalamic–pituitary–adrenal cortical system occurs only with very high doses of aspirin. The effects on thyroid function may be seen with lower doses but are of minimal clinical significance. Thus *in vitro* studies have demonstrated that salicylates may displace thyroid hormones from plasma protein-binding sites [113] and suppress production of thyroid-stimulating hormone by the pituitary and hypothalamus [114].

AUDITORY, VESTIBULAR AND CENTRAL NERVOUS SYSTEM EFFECTS

In low therapeutic doses, aspirin appears to cause no central nervous system symptoms or signs, but higher doses characteristically result in tinnitus and hearing loss. These are regarded as the first signs of salicylate toxicity or salicylism and tend to emerge with plasma salicylate concentrations above 200 mg/l (1.44 mmol/l). They are clearly dose related and are fully reversible, resolving within a few days of stopping the drug [115]. Retrospective drug surveillance studies suggested that larger doses of aspirin given at longer intervals of time caused a higher incidence of deafness than similar total daily doses divided into smaller individual doses given more frequently. This presumably reflects differences in peak plasma salicylate concentrations [116]. The underlying mechanism is uncertain as histopathological studies have failed to identify any relevant changes. It is thought that there may be increased labyrinthine pressure or an effect on the hair cells of the cochlea, possibly secondary to vasoconstriction of the microvasculature of the inner ear [117].

High-dose chronic aspirin therapy and overdose are associated with a range of CNS symptoms, the characteristic accompanying acidosis resulting in easier transfer of salicylate across the blood–brain barrier. Symptoms of irritability, tremor, confusion, psychosis, dizziness and convulsions may be followed by drowsiness, stupor and eventually coma.

Recently, interest in possible neuroprotective effects of aspirin therapy has arisen from clinical and experimental evidence suggesting that neurodegeneration is often associated with inflammation [118] and from epidemiological studies demonstrating an inverse relationship between non-steroidal anti-inflammatory drug therapy and onset of Alzheimer's disease [119]. In laboratory studies, anti-inflammatory serum concentrations of aspirin were 83% protective against neurotoxicity elicited by the excitatory amino acid glutamate in rat primary neuronal cultures and helped to preserve cell viability in hippocampal slices [120]. This appeared to be due to inhibition of the glutamate-mediated induction of nuclear factor kappa B, rather than directly on the glutamate receptors themselves and is an action which is independent of COX inhibition and not shared with other non-steroidal anti-inflammatory drugs.

Conclusions

Prostaglandins are implicated in many physiological and pathological processes throughout the body. The irreversible acetylation of COX by aspirin might be expected to have major effects, both from failure to synthesize eicosanoids such as thromboxane and prostacyclin and from the shift towards the lipoxygenase pathway of arachidonic acid metabolism. Other pharmacological and biochemical effects probably contribute fairly little to most of aspirin's therapeutic properties, but the search for alternative explanations continues.

KEYPOINTS

- Aspirin inhibits cyclo-oxygenase (COX), a key enzyme in prostaglandin synthesis, through acetylation of the amino acid serine in the active centre. This is considered the main mechanism responsible for most of its pharmacological effects.
- Other actions include effects on leucocyte function, free radical scavenging properties, lipoxygenase activity, cell membrane processes and various other endocrine and immunological processes. These may be especially relevant to aspirin's anti-inflammatory effects.
- Inhibition of platelet COX-1 results in decreased production of the proaggregatory prostaglandin thromboxane A2. The antiplatelet effect is permanent, unlike the reversible inhibition of endothelial COX-1, which temporarily reduces prostacyclin production. The differential sensitivity of platelet and endothelial enzymes is dose dependent and accounts for aspirin's antithrombotic properties.
- The metabolic, renal, hepatic, cardiovascular and CNS effects of aspirin only occur with high therapeutic doses or after overdose.

REFERENCES

1. Smith MJH, Smith PK. The Salicylates. A Critical Bibliographic Review. New York: Interscience Publishers, 1966.
2. Vane JR. Inhibition of prostaglandin synthesis as a mechanism of action of aspirin-like drugs. Nature 1971; 231: 232–235.
3. Ferreira SH. Prostaglandins, aspirin-like drugs and analgesia. Nature 1972; 240: 200–203.
4. Roth GJ, Majerus PW. The mechanism of the effect of aspirin on human platelets. 1. Acetylation of a particulate fraction protein. J Clin Invest 1975; 56: 624–632.

5. Smith WL. Prostanoid biosynthesis and mechanisms of action. Am J Physiol 1992; 263: F181–F191.
6. Fu JY, Masferrer JL, Siebert K, Raz A, Needleman P. The induction and suppression of prostaglandin H2 synthetase (cyclooxygenase in human monocytes). J Biol Chem 1990; 265: 16737–16740.
7. Meade EA, Smith WL, DeWitt DL. Differential inhibition of prostaglandin endoperoxide synthase (cyclooxygenase) isoenzymes by aspirin and other nonsteroidal antiinflammatory drugs. J Biol Chem 1993; 268: 6610–6614.
8. Loll PJ, Picot D, Garavito RM. The structural basis of aspirin activity inferred from the crystal structure of inactivated prostaglandin H2 synthase. Nature Struct Biol 1995; 2: 637–643.
9. Allen KN. Aspirin – now we can see it. Nature Med 1995; 9: 882–883.
10. Vane JR, Flower RJ, Botting RM. History of aspirin and its mechanism of action. Stroke 1990; 21(suppl IV): IV-12–IV-23.
11. Atkinson D, Collier H. Salicylates: molecular mechanisms and therapeutic action. Adv Pharmacol Chemother 1981; 17: 233–238.
12. Abramson SB, Weissmann G. The mechanisms of action of non-steroidal anti-inflammatory drugs. Arthritis Rheum 1989; 32: 1–9.
13. Forrest M, Brooks PM. Mechanism of action of non-steroidal anti-rheumatic drugs. Baillière's Clin Rheumatol 1988; 2: 275–294.
14. Higgs GA, Salmon JA, Henderson B, Vane JR. Pharmacokinetics of aspirin and salicylate in relation to inhibition of arachidonate cyclooxygenase and anti-inflammatory activity. Proc Natl Acad Sci USA 1987; 84: 1417–1420.
15. Higgs GA, Salmon JA. Cyclooxygenase products in carrageenin-induced inflammation. Prostaglandins 1979; 17: 737–746.
16. Trang LE, Grantstrom E, Lovgren O. Levels of prostaglandin F2 alpha and thromboxane B2 in joint fluid in rheumatoid arthritis. Scand J Rheumatol 1977; 6: 151–154.
17. Brune K, Graf P. Non-steroid anti-inflammatory drugs: influence of extracellular pH on biodistribution and pharmacological effects. Biochem Pharmacol 1978; 27: 525–526.
18. Smith MJ. Aspirin and prostaglandins: some recent developments. Agents Actions 1978; 8: 427–429.
19. Smith MJH, Ford-Hutchinson AW, Walker JR, Slack JA. Aspirin, salicylate and prostaglandins. Agents Actions 1979; 9: 483–487.
20. Blechman WJ, Lechner BL. Clinical comparative evaluation of choline magnesium trisalicylate and acetylsalicylic acid in rheumatoid arthritis. Rheumat Rehab 1979; 18: 119–124.
21. Preston SJ, Arnold MH, Beller EM, Brooks PM, Buchanan WW. Comparative analgesic and anti-inflammatory properties of sodium salicylate and acetylsalicylic acid (aspirin) in rheumatoid arthritis. Br J Clin Pharmacol 1989; 27: 607–611.
22. Henderson B, Higgs GA, Salmon JJ. Is aspirin a pro-drug for salicylate? Br J Clin Pharmacol 1986; 88 (suppl): 400–405.
23. Paulus HE. Aspirin versus nonacetylated salicylate. J Rheumatol 1989; 16: 264–265.

24. April P, Abeles M, Baraf H *et al.* Does the acetyl group in aspirin contribute to the anti-inflammatory efficacy of salicylic acid in the treatment of rheumatoid arthritis? Semin Arthritis Rheum 1990; 19 (suppl 2): 20–28.
25. McArthur JN, Dawkins PD, Smith MJH, Hamilton EB. Mode of action of antirheumatic drugs. Br Med J 1971; 2: 677–679.
26. Walker JR, Smith MJH, Ford-Hutchinson AW. Anti-inflammatory drugs, prostaglandins and leucocyte migration. Agents Actions 1976; 6: 602–606.
27. Bonta IL, Bult H, Vincent JE, Zijlstra FJ. Acute anti-inflammatory effects of aspirin and dexamethasone in rats deprived of endogenous prostanglandin precursors. J Pharm Pharmacol 1977; 29: 1–7.
28. Minta JO, Williams MD. Some nonsteroidal antiinflammatory drugs inhibit the generation of superoxide anions by activated polymorphs by blocking ligand–receptor interactions. J Rheumatol 1985; 12: 751–757.
29. Udassin R, Ariel I, Haskel Y, Kitrossky N, Chevion M. Salicylate as an in vivo free radical trap: studies on ischaemic insult to the rat intestine. Free Radic Biol Med 1991; 10: 1–6.
30. Seigel MI, McConnell RT, Cuatrecasas P. Aspirin-like drugs interfere with arachidonate metabolism by inhibition of the 12-hydroperoxy-5,8,10,14-eicosatetraenoic acid peroxidase activity of the lipoxygenase pathway. Proc Natl Acad Sci USA 1979; 76: 3774–3778.
31. Bray MA. Leukotrienes in inflammation. Agents Actions 1986; 19: 87–99.
32. Voigtlander V, Hansch GM, Rother U. Effect of aspirin on complement in vivo. Int Arch Allergy Appl Immunol 1980; 61: 150–158.
33. Glenn EM, Bowman BJ, Rohloff N. Anomalous biological effects of salicylates and prostaglandins. Agents Actions 1979; 9: 257–264.
34. Bomalaski JS, Hirata F, Clark MA. Aspirin inhibits phospholipase C. Biochem Biophys Res Commun 1986; 139: 115–121.
35. Ignarro LJ. Lysosome membrane stabilization in vivo: effect of steroidal and nonsteroidal antiinflammatory drugs on the integrity of rat liver lysosomes. J Pharmacol Exp Ther 1972; 182: 179–188.
36. Weissmann G. Aspirin. Scientific American 1991; 26: 484–490.
37. Spector WG, Willoughby DA. Anti-inflammatory effects of salicylate in the rat. In: Dixon A St J, Martin BK, Smith MJH, Wood PHN (eds) Salicylates: An International Symposium. Boston: Little, Brown and Co., 1963: 141–147.
38. McKenzie LS, Horsburgh BA, Ghosh P, Taylor TKF. Effect of anti-inflammatory drugs on sulphated glycosaminoglycan synthesis in aged human articular cartilage. Ann Rheum Dis 1976; 35: 487–497.
39. Lim RKS. Salicylate analgesia. In: Smith MJH, Smith PK (eds) The Salicylates: A Critical Bibliographic Review. New York: Interscience, 1966: 155–188.
40. Sunshine A, Olson NZ. Nonnarcotic analgesics. In: Wall PD, Melzack R (eds) Textbook of Pain. Edinburgh: Churchill Livingstone, 1994: 923–942.
41. Lim RKS, Guzman F, Rodgers DW *et al.* Site of action of narcotic and non-narcotic analgesics determined by blocking bradykinin-evoked visceral pain. Arch Int Pharmacodyn 1964; 152: 25–58.

42. Scott D Jr. Aspirin action on receptors in the tooth. Science 1968; 161: 180–181.
43. Gilfoil TM, Klavins I, Grumbach L. Effects of acetylsalicylic acid on the oedema and hyperesthesia of the experimentally inflamed rat's paw. J Pharmacol Exp Ther 1963; 142: 1–5.
44. Clissold SP. Aspirin and related derivatives of salicylic acid. Drugs 1986; 32 (suppl 4): 8–26.
45. Schild HO. Applied Pharmacology: Analgesic Drugs. Edinburgh: Churchill Livingstone, 1980: 278–294.
46. Seymour RA, Rawlins MD. The efficacy of pharmacokinetics of aspirin in post-operative dental pain. Br J Clin Pharmacol 1982; 13: 807–810.
47. Velagapundi R, Harter JG, Brueckner R, Peck CC. Pharmacokinetic/pharmacodynamic models in analgesic study design. In: Porteno R, Laska E (eds) Advances in Pain Research and Therapy, Vol 18. New York: Raven Press, 1991: 559–562.
48. Seymour RA, Williams FM, Ward A, Rawlins MD. Aspirin metabolism and efficacy in postoperative dental pain. Br J Clin Pharmacol 1984; 17: 697–702.
49. Williams FM, Wynne H, Woodhouse KW, Rawlins MD. Plasma aspirin esterase: the influence of old age and frailty. Age Ageing 1989; 18: 39–42.
50. Malmberg AB, Yaksh TL. Hyperalgesia mediated by spinal glutamate or substance receptor blocked by spinal cyclo-oxygenase inhibition. Science 1992: 257: 1277–1280.
51. Ferreira SH, Lorenzetti B, Correa FMA. Central and peripheral antialgesic actions of aspirin-like drugs. Eur J Pharmacol 1978; 53: 39–48.
52. Yaksh TL. Central and peripheral mechanisms for the analgesic action of acetylsalicylic acid. In: Barnett HJM, Hirsh, J, Mustard JF (eds) Acetylsalicylic Acid: New Uses for an Old Drug. New York: Raven Press, 1982: 137–151.
53. Cooper KE, Veale WL, Kasting NW. Temperature regulation, fever and antipyretics. In Barnett HJM, Hirsh J, Mustard JF (eds) Acetylsalicylic Acid: New Uses for an Old Drug. New York: Raven Press, 1982: 153–160.
54. Walson PD. Fever. In: Rakel RE (ed.) Conn's Current Therapy. Philadelphia: WB Saunders, 1997: 24–27.
55. Milton AS. Modern views on the pathogenesis of fever and the mode of action of antipyretic drugs. J Pharm Pharmacol 1976; 28: 393–399.
56. Milton AS. Prostaglandin E1 and endotoxin fever and the effects of aspirin, indomethacin and 4-acetamidophenol. Adv Biosci 1973; 9: 495–500.
57. Seed JC. A clinical comparison of the antipyretic potency of aspirin and sodium salicylate. Clin Pharmacol Ther 1965; 6: 354–358.
58. Jacobson ED, Bass DE. Effects of sodium salicylate on physiological response to work in heat. J Appl Physiol 1964; 19: 33–36.
59. Fred HL. The 100-mile run: preparation, performance, and recovery. A case report. Am J Sports Med 1981; 9: 258–261.
60. O'Brien JR. Effects of salicylates on human platelets. Lancet 1988; 1: 779–781.

61. Patrono C. Aspirin as an antiplatelet drug. N Engl J Med 1994: 330: 1287–1294.
62. Link KP, Overman RS, Sullivan WR, Huebner CF, Scheel LD. Studies of the hemorrhagic sweet clover disease. XI. Hypoprothrombinaemia in the rat induced by salicylic acid. J Biol Chem 1943; 147: 463–474.
63. Park BK, Leck JB. On the mechanism of salicylate-induced hypothrombinaemia. J Pharm Pharmacol 1981; 33: 25–28.
64. Menon SI. Aspirin and blood fibrinolysis. Lancet 1970; 1: 364.
65. Moroz LA. Increased blood fibrinolytic activity after aspirin ingestion. N Engl J Med 1977; 296: 525–529.
66. Pedersen AK, FitzGerald GA. Dose-related kinetics of aspirin: presystemic acetylation of platelet cyclooxygenase. N Engl J Med 1984; 311: 1206–1211.
67. Weiss JH, Aledorf LM, Kochwa S. The effect of salicylates on the haemostatic properties of platelets in man. J Clin Invest 1968; 47: 2169–2173.
68. Seymour RA, Williams FM, Oxley A *et al.* A comparative study of the effects of aspirin and paracetamol on platelet aggregation and bleeding time. Eur J Clin Pharmacol 1984; 26: 567–572.
69. Patrignani P, Filabozzi P, Patrono C. Selective cumulative inhibition of platelet thromboxane production by low-dose aspirin in healthy subjects. J Clin Invest 1982; 69: 1366–1372.
70. Weksler BB, Pett SB, Alonso D *et al.* Differential inhibition by aspirin of vascular and platelet prostaglandin synthesis in atherosclerotic patients. N Engl J Med 1983; 308: 800–805.
71. Fiore LD, Brophy MT, Lopez A, Janson P, Deykin D. The bleeding time response to aspirin. Identifying the hyperresponder. Am J Clin Pathol 1990; 94: 292–296.
72. Ghooi RB, Thatte SM, Joshi PS. The mechanism of action of aspirin – is there anything beyond cyclo-oxygenase? Med Hypoth 1995; 44: 77–80.
73. Bjornsson CP, Schneider DE, Berger H Jr. Aspirin acetylates fibrinogen and enhances fibrinolysis: fibrinolytic effect is independent of changes in plasminogen activator levels. J Pharmacol Exp Ther 1989; 250: 154–161.
74. Ezratty AM, Simon DI, Loscalzo J. Acetylated fibrinogen facilitates plasminogen activation and attenuates platelet aggregation. Clin Res 1992; 40: 201.
75. Szczeklil A, Krzanowski M, Gora P, Radwan J. Antiplatelet drugs and generation of thrombin in clotting blood. Blood 1992; 80: 2006–2011.
76. Husted SE, Nielsen HK, Petersen T, Mogensen CE, Geday E. Acute effects of acetylsalicylic acid on renal and hepatic function in normal humans. Int J Clin Pharmacol Ther Toxicol 1985; 23: 141–144.
77. FitzGerald GA, Oates JA, Hawiger J *et al.* Endogenous biosynthesis of prostacyclin and thromboxane and platelet function during chronic administration of aspirin in man. J Clin Invest 1983; 71: 676–688.
78. Brezinski DA, Nesto RW, Serhan CN. Angioplasty triggers intracoronary leukotrienes and lipoxin A2; impact of aspirin therapy. Circulation 1992; 86: 56–63.

79. Oates JA, Fitzgerald GA, Branch RA *et al.* Clinical implications of prostaglandin and thromboxane A2 formation. N Engl J Med 1988; 319: 689–698, 761–767.
80. Sullivan JL. Gastric safety and enteric-coated aspirin. Lancet 1997; 349: 431–432.
81. Sullivan JL. Iron vs cholesterol. Perspectives on the iron and heart disease debate. J Clin Epidemiol 1996; 49: 1345–1352.
82. Stevens RG, Graubaard BI, Micozzi MS, Neriishi K, Blumberg BS. Moderate elevation of body iron level and increased risk of cancer occurrence and death. Int J Cancer 1994; 53: 364–369.
83. Vane JR. Aspirin. In: Breckenridge A (ed.) Advanced Medicine. Topics in Therapeutics 1. London: Pitman Medical, 1975.
84. Robert A. Cytoprotection by prostaglandins. Gastroenterology 1979; 77: 761–767.
85. Lichtenstein DR, Syngal S, Wolfe MM. Nonsteroidal antiinflammatory drugs and the gastrointestinal tract. Arthritis Rheum 1995; 38: 5–18.
86. Davenport HW. Salicylate damage to the gastric mucosal barrier. N Engl J Med 1967; 276: 1307–1309.
87. Cohen MM, Pollett JM. Prostaglandin E2 prevents aspirin and indomethacin damage to human gastric mucosa. Surg Forum 1976; 27: 400–401.
88. Kauffman GL, Grossman MI. Prostaglandin and cimetidine inhibit the formation of ulcers produced by parenteral salicylates. Gastroenterology 1978; 75: 1099–1102.
89. Goddard AF, Donnelly MT, Filipowicz B *et al.* Low dose misoprostol as prophylaxis against low-dose aspirin-induced gastroduodenal mucosal injury. Gut 1996; 39: A33.
90. Roth S, Agarwal N, Mahowald M *et al.* Misoprostol heals the gastroduodenal injury in patients with rheumatoid arthritis receiving aspirin. Arch Intern Med 1989; 149: 775–779.
91. Rainsford KD. Gastrointestinal and other side effects from the use of aspirin and related drugs: biochemical studies on the mechanisms of gastrotoxicity. Agents Actions 1977; 1 (suppl): 59–70.
92. McDonald JWD. Effects of acetylsalicylic acid on gastric mucosa. In: Barnett HJM, Hirsh J, Mustard JF (eds) Acetylsalicylic Acid: New Uses for an Old Drug. New York: Raven Press, 1982: 87–94.
93. Anonymous. Which prophylactic aspirin? Drugs Therapeut Bull 1997; 35: 7–8.
94. Lichtenberger LM, Wang Z, Romero JJ *et al.* Non-steroidal anti-inflammatory drugs (NSAIDs) associate with zwitterionic phospholipids: insight into the mechanism and reversal of NSAID induced gastrointestinal injury. Nature Med 1995; 1: 154–158.
95. Insel PA. Analgesic–antipyretic and antiinflammatory agents: the salicylates. In Hardman JG, Limbird LE (eds) Goodman and Gilman's The Pharmacological Basis of Therapeutics, 9th edn. New York: Mosby, 1996: 617–657.

96. Muther RS, Potter DM, Bennett WM. Aspirin-induced depression of glomerular filtration rate in normal humans: role of sodium balance. Ann Intern Med 1981; 94: 317–321.
97. Plotz PH, Kimberly RP. Acute effects of aspirin and acetaminophen on renal function. Arch Intern Med 1981; 141: 343–348.
98. Cooper K, Bennett W. Nephrotoxicity of common drugs used in clinical practice. Arch Intern Med 1987; 147: 1213–1218.
99. Perneger TV, Whelton PK, Klag MJ. Risk of kidney failure associated with the use of acetaminophen, aspirin and nonsteroidal anti-inflammatory drugs. N Engl J Med 1994; 331: 1675–1679.
100. Prescott LF. Analgesic nephropathy. Drugs 1982; 23: 75–149.
101. Rainsford KD. Aspirin and the Salicylates. London: Butterworths, 1984.
102. Smith MJH. Metabolic effects of salicylates. In: Smith MJH, Smith PK (eds) The Salicylates. A Critical Bibliographic Review. New York: Interscience Publishers, 1966: 107–153.
103. Pinckard RN, Hawkins D, Farr RS. In vitro acetylation of plasma proteins, enzymes and DNA by aspirin. Nature 1968; 219: 68–69.
104. Smith MJH, Dawkins PD. Salicylate and enzymes. J Pharm Pharmacol 1971; 23: 729–744.
105. Smith MJH. Interactions with endocrine systems. In: Smith MJH, Smith PK (eds) The Salicylates. A Critical Bibliographic Review. New York: Interscience Publishers, 1966: 49–105.
106. Adams SS, Cobb R. A possible basis for anti-inflammatory activity of salicylates and other non-hormonal anti-rheumatic drugs. Nature 1958; 181: 773–744.
107. Giugliano D, Torella R, Siniscalchi N, Improta L, D'Onofrio F. The effect of acetylsalicylic acid on insulin response to glucose and arginine in normal man. Diabetologia 1978; 14: 359–362.
108. Fang VS, Foye WO, Robinson SM, Jenkins HJ. Hypoglycemic activity and chemical structure of the salicylates. J Pharmaceut Sci 1968; 57: 2111–2116.
109. Gaitonde BB, Joglekar SN, Shaligram SV. The hyperglycaemic action of sodium salicylate. Br J Pharmacol Chemother 1967; 30: 554–560.
110. Seltzer HS. Drug-induced hypoglycaemia: a review of 1418 cases. Endocrinol Metab Clin North Am 1989; 18: 163–183.
111. Pandit MK, Burke J, Gustafson AB, Minocha A, Peiris AN. Drug-induced disorders of glucose tolerance. Ann Intern Med 1993; 118: 529–539.
112. Reid J, Lightbody TD. The insulin equivalence of salicylate. Br Med J 1959; 1: 897–900.
113. Christensen LL. Thyroxine-releasing effect of salicylate and of 2,4-dinitrophenol. Nature 1959; 183: 1189–1190.
114. Wolff J, Austen FK. Salicylates and thyroid function. II. The effect on the thyroid–pituitary interrelation. J Clin Invest 1958; 37: 1144–1152.
115. Jick H. Adverse effects of acetylsalicylic acid. In: Barnett HJM, Hirsh J, Mustard JF (eds) Acetylsalicylic Acid: New Uses for an Old Drug. New York: Raven Press, 1982: 239–247.

116. Miller RR. Deafness due to plain and long-acting aspirin tablets. J Clin Pharmacol 1978; 18: 468–471.
117. McPherson DL, Miller JM. Choline salicylate. Effects on cochlear function. Arch Otolaryngology 1974; 99: 304–308.
118. McGeer PL, McGeer EG. The inflammatory response system of brain implications for therapy of Alzheimer's and other neurodegenerative diseases. Brain Res Rev 1995; 21: 195–218.
119. Stewart WF, Kawas C, Corrada M, Metter EJ. Risk of Alzheimer's disease and duration of NSAID use. Neurology 1997; 48; 626–632.
120. Grilli M, Pizzi M, Memo M, Spano P. Neuroprotection by aspirin and sodium salicylate through blockade of NF-kB activation. Science 1996; 274: 1383–1385.
121. Meredith TJ, Vale JA, Proudfoot, AT. Poisoning caused by analgesic drugs. In: Weatherall DJ, Ledingham JGG, Warrell DA (eds) Oxford: University Press, 1996: 1051–1056.

Therapeutic use for inflammation, pain and fever

Introduction

Some 2000 years ago, the Roman physician Celsus described the four cardinal signs of inflammation – rubor, calor, dolor, tumor (redness, heat, pain and swelling). For much of the 20th century, aspirin was the mainstay of treatment for most inflammatory diseases, for everyday management of pain and discomfort and for the symptomatic treatment of fever. In the last 30 years, the place of aspirin as the first-choice treatment for most rheumatic conditions has been taken over by a wide choice of other newer non-steroidal anti-inflammatory drugs. Similarly, its pre-eminent position as the analgesic of choice for mild to moderate pain has been challenged by the introduction of many non-narcotic analgesic drugs and its antipyretic use has become rare, especially in children, because of concerns about drug-induced Reye's syndrome and the availability of safer alternatives. Despite its more limited clinical use for these indications, aspirin remains the standard for the comparison and evaluation of other newer anti-inflammatory, analgesic and antipyretic agents. It is still the common household painkiller, especially among older people. In a survey conducted in the early 1990s of the use of over-the-counter (non-prescription) medication by older people in America, aspirin was reported to have been taken within the previous two weeks by 42% of the study population. In nearly all cases this was apparently being taken for its analgesic properties. No other drug was used more often [1]. Similar findings have previously been reported from studies in Britain [2] and New Zealand [3].

Inflammation

Aspirin provides symptomatic relief of pain and stiffness in inflammatory arthritis, though it does not alter the natural history of disease. Its use has now been largely replaced by newer non-steroidal anti-inflammatory drugs which may be better tolerated [4] but are considerably more expensive and not significantly more effective. Reported differences in efficacy appear to relate more to the comparative dosages used than to the particular anti-inflammatory properties of any specific drug.

Muscle pain and soft tissue injury associated with trauma or sports injuries will respond well to aspirin, though it is unproven whether it has advantages over simple analgesics such as paracetamol and sometimes may not be tolerated as well as other non-steroidal anti-inflammatory drugs.

It is generally accepted that plasma concentrations of salicylate of at least 150 mg/ml (1.1 mmol/l) are necessary to achieve adequate anti-inflammatory activity. However, concentrations above 300 mg/ml (2.2 mmol/l) are often associated with unacceptable and potentially dangerous adverse systemic effects and so the therapeutic window for the anti-inflammatory use of aspirin is relatively narrow [5]. The aim must be to use the lowest effective dose, which is often the highest tolerated. In the past, many clinicians first established the dose of aspirin which caused tinnitus and then slightly reduced this to a maintenance 'subtinnitus level'. A more satisfactory approach is to use plasma salicylate measurement (usually best taken at the 'trough' point in the dosing interval, just before the next dose) as an adjunct to clinical assessment, particularly in the evaluation of inadequate disease control or possible toxicity [6].

The usual anti-inflammatory dose of aspirin is 600–900 mg every 4–6 hours. At this dosage, steady state concentrations will be reached after about four or five days. Both the dose size and dosing interval can be flexible, provided that the total daily dosage is adequate. This should not exceed 4000 mg. One review suggested that a rational anti-inflammatory aspirin dosage regimen was to use 60 mg/kg per day in one to six divided doses [7]. Certainly, given the long half life of salicylate at these doses, intervals of eight or even 12 hours between dosing, rather than the traditional 4–6 hours, are usually sufficient to maintain peak and trough plasma salicylate concentrations in the anti-inflammatory range [8]. Thus twice-daily dosing was shown to be effective in both children with juvenile rheumatoid arthritis [9] and in adults with a range of rheumatic diseases [10]. There is, however, large interpatient variability and final dosage regimens should be decided on clinical grounds, perhaps guided by plasma salicylate measurement.

RHEUMATOID ARTHRITIS

Many studies conducted in the 1960s and 1970s confirmed that high-dose aspirin can rapidly and effectively suppress the inflammation of rheumatoid arthritis, but does not significantly slow disease progression or limit joint damage [11–13]. It is superior to placebo or paracetamol [14] and was for many years the mainstay of treatment [15]. It has now been largely superseded by the newer non-steroidal anti-inflammatory agents, which have the advantage of better tolerance and generally a need for less frequent dosing of a smaller bulk of drug and hence better compliance. Aspirin, often in suboptimal dosage, was generally used as the comparator in early clinical trials of these agents and was generally at least as effective, but associated with more adverse effects [16].

For the small minority of patients who continue to be satisfied with the symptomatic effects of aspirin, there seems little reason to change to another agent. In general, however, the poor tolerance of the high doses of aspirin which are often required in this indication has meant that it has been almost totally replaced by other non-steroidal anti-inflammatory drugs. Some recent data suggest that in the doses actually used in clinical practice, aspirin may be among the safest of non-steroidal anti-inflammatory drugs and it merits reconsideration as adjunctive therapy for management of rheumatoid arthritis [17].

JUVENILE RHEUMATOID ARTHRITIS (STILL'S DISEASE)

Since the recognition of the importance of aspirin as a possible risk factor for the development of Reye's syndrome, Still's disease is now one of the few acceptable indications for the use of aspirin in children [18]. Many authorities still regard it as standard treatment. Certainly aspirin is at least as effective as other non-steroidal anti-inflammatory drugs and is used in doses up to 120 mg/kg/day in order to maintain plasma salicylate concentrations in the high anti-inflammatory range [19]. This has been shown to produce the best clinical response [20] but is also very close to the toxic range. Children typically do not complain of the early warning signs of salicylate toxicity, such as tinnitus and mild deafness, and regular plasma monitoring to guide dosage is therefore essential [21]. This will also serve as a useful check on compliance, which is probably a more common clinical problem than the effects of overdosage.

OSTEOARTHRITIS

Aspirin is still often used for self-medication of pain and stiffness associated with osteoarthritis and is generally effective. Comparative studies with non-steroidal anti-inflammatory drugs such as naproxen [22] and diflunisal [23] have shown similar efficacy, but more gastrointestinal intolerance. Concern about deleterious effects of potent prostaglandin synthetase inhibitors on cartilage metabolism and the synthesis of proteoglycans by chondrocytes both *in vitro* and *in vivo* [24,25] suggests that the chronic use of aspirin for this indication should be avoided. In animal models, effects on cartilage metabolism were more marked in osteoarthritic than in normal joints [26].

OTHER ARTHRITIC CONDITIONS

Aspirin is still occasionally used by patients with other inflammatory rheumatic conditions, but is generally poorly effective. Exceptions may be systemic lupus erythematosus, where high-dose aspirin was superior to ibuprofen in controlling joint pain [27], and in particular antiphospholipid syndrome, where low-dose aspirin may be useful in controlling thrombotic complications [28]. In patients with ankylosing spondylitis, a crossover comparative trial of aspirin 4 g daily, indomethacin 100 mg daily and phenylbutazone 300 mg daily found that aspirin was significantly less effective than the comparators [29].

Despite the beneficial effects of high-dose aspirin on enhancing urinary excretion of urate, aspirin is rarely used to treat acute gout. In low dose, aspirin inhibits renal tubular secretion of urate and leads to uric acid retention, increasing the risk of precipitating an attack [30].

BARTTER'S SYNDROME

This rare inherited disorder is characterized by the increased synthesis of prostaglandins, leading to increased renin and aldosterone activity and so to hypokalaemia, muscle weakness, cramps and growth disturbance [31]. There is also increased production of bradykinin. The syndrome usually presents in early childhood. Treatment with aspirin in a dose of 50–100 mg/kg body weight effectively suppresses prostaglandin production and significantly improves symptoms and enhances growth and development [32,33].

Pain

Aspirin is useful for treating mild to moderate pain and is included in Step 1 of the World Health Organization's 'analgesic ladder' and together with an opioid (such as codeine) for more severe pain in Step 2 [34]. It is most effective in the treatment of deep somatic pain arising from integumental structures of muscles, joints, tendons and fasciae, in headache resulting from distension of blood vessels and meninges and in pain originating from nerve trunks, especially when associated with inflammation and swelling. It is relatively ineffective in treating superficial, cutaneous pain and is ineffective in deep, visceral pain [35].

In the British National Formulary [36], aspirin is recommended as the analgesic of choice for headache, transient musculoskeletal aches and pains and dysmenorrhoea and may also be useful in managing toothache, discomfort associated with mild viral illness and postsurgical and

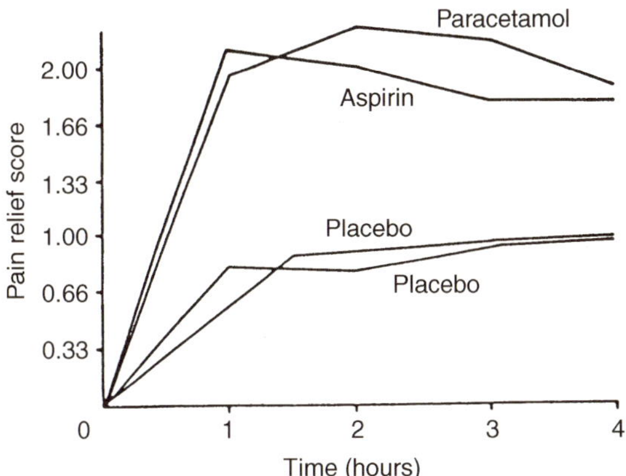

Fig. 4.1
Composite time–pain relief curves from two separate studies comparing aspirin 650 mg with placebo and paracetamol 650 mg with placebo in patients following oral surgery (from reference [40] with permission).

postpartum/episiotomy pain. Worldwide, aspirin is still more widely used for pain relief than any other single drug. It has the advantage over paracetamol of having anti-inflammatory as well as analgesic activity and does not have the problems of tolerance and physical dependence which are associated with narcotic analgesics.

In most clinical trials where aspirin has been compared with placebo, the active drug has nearly always produced significantly greater pain relief. Many such studies have been conducted in patients with postoperative dental pain [37,38] or headache [39]. Comparisons with paracetamol have concluded that the two drugs are equianalgesic and, on a weight-to-weight basis, are equipotent in most types of pain, with similar dose–response and time–effect curves [40,41] (Figure 4.1). However, no systematic review or meta-analysis has yet been published. Comparisons with codeine have demonstrated aspirin's analgesic superiority when the drugs are given alone, although the combination provides the most effective pain relief [42].

The standard analgesic dose of aspirin is 600–1200 mg, taken every 4–6 hours when necessary. The analgesic activity of aspirin and the duration of its effect increase with increasing dose up to 1200 mg [43–45]. Increasing the dose beyond this level does not produce any additional benefit. Nor is there much evidence of increasing analgesic effect with repeated dosing. This suggests that salicylic acid, which will accumulate with repeated dosing, contributes relatively little to analgesia compared to aspirin itself, which is always rapidly metabolized [8]. Thus a second 650 mg aspirin dose administered to postpartum patients four hours after their first dose had little if any additional analgesic effect [46] and there was only a slight increase in reported analgesic effect of aspirin in male patients with chronic pain during six days of drug administration [47]. The analgesic efficacy of aspirin in any particular patient can thus be evaluated within a short time and adjuvant analgesia added or an alternative drug substituted if appropriate.

More rapidly absorbed preparations might be expected to result in earlier and greater pain relief [48], although clinical evidence to support this is sparse. Sustained-release preparations would not be expected to provide prolonged analgesia and the duration of pain relief produced by regular and sustained-release aspirin has been found to be essentially identical [49].

MIGRAINE AND HEADACHE

Aspirin has long been used to treat acute migraine and is significantly better than placebo, with relief rates of between 34% and 45% [50,51]. There is only limited evidence from double-blind controlled trials of use of aspirin in migraine prophylaxis and the optimal dose and efficacy are uncertain. Originally it was thought that platelet aggregation might play a central role in causing migraine [52,53], thus justifying use of aspirin, although this is now thought to be less important. Nevertheless, a minority of patients seem to respond well to long-term prophylaxis [54,55] and there was a significant reduction in reported migraine attacks in subjects taking low-dose aspirin in the large primary prevention studies of American [56] and British doctors [57].

In the treatment of acute attacks, high doses (1200 mg) are often required and a second dose taken two hours after an ineffective first dose may be useful [58]. Dispersible or effervescent preparations are to be preferred as they will be more rapidly absorbed and therefore give earlier pain relief. Gastric status, prolonged gastric emptying and the nausea and vomiting which often accompany attacks all slow down absorption. Thus plasma salicylate concentrations 30 and 60 minutes after oral administration of 900 mg of effervescent aspirin were significantly lower in patients during a migraine attack than when the same aspirin dose was given to the same patients when migraine free [59]. The delay in absorption may be avoided by predosing with metoclopramide [60] and the combination has been shown to be more effective than aspirin alone [50], although not as effective as oral sumatriptan [51]. Unfortunately the only aspirin–metoclopramide combination product available in Britain has recently been withdrawn. Domperidone may be preferable in elderly patients and in those aged under 20 years, who are at greater risk of metoclopramide-associated dystonic reactions.

Episodic non-migrainous or tension-type headache, which is accompanied by little or no nausea and no aura symptoms, may also respond well to aspirin. It should not be taken daily for more than a day or two. Regular use is to be discouraged because of the risk of development of chronic headache induced by analgesic abuse [61]. Whilst more common with codeine-based analgesics, the phenomenon has also been attributed to aspirin when doses greater than 50 g a month have been regularly taken for three months or more. It has only been described when analgesics have been taken for headaches and not when they have been taken for other painful conditions such as low back pain. The situation is exacerbated by the fact that analgesic abuse seems to nullify the effects of prophylactic drugs given for the original headache condition [62]. The typical patient is female, with a long history of taking escalating doses of analgesic drugs which provide increasingly less effective pain relief, has an addictive personality and a tendency towards tranquillizer abuse [63].

DYSMENORRHOEA

Increased prostaglandin production is assumed to be an important factor associated with dysmenorrhoea [64]. It therefore seems reasonable to expect aspirin to be particularly useful in management and the British National Formulary specifically recommends its use. Limited clinical trial evidence of its superiority to placebo is generally supportive [65], though regular use of relatively high doses (e.g. 600 mg four times daily) may be required. However, there are some reports that it is less effective than other non-steroidal anti-inflammatory drugs [66], possibly because most of the drug reaching the uterus has already undergone hydrolysis and therefore acetylation of uterine COX is not possible [67].

HERPES ZOSTER AND NEURALGIA

Topical application of aspirin suspended in chloroform or diethyl ether has been reported to be highly effective in relieving the pain of acute herpes zoster and of postherpetic neuralgia [68–70]. In a crossover comparison with two other non-steroidal anti-inflammatory drug/ether mixtures, only the aspirin preparation was significantly superior to placebo, with good-to-excellent results being achieved in over 80% of patients. Patients with trigeminal involvement, with less severe pain and with dysaesthetic quality of pain were said to respond best [70]. Benefit from topical application of aspirin in chloroform has also been claimed in the management of chronic neurogenic pain in patients in whom conventional methods of treatment had been unsuccessful [71].

Fever

A raised temperature is a common manifestation of many medical conditions, but only when very high (> 41°C) may it cause harm. Nevertheless, fever is the single most common symptom for which patients see their general practitioner and many will take antipyretic medication [72]. Until a few years ago, aspirin was widely used as an antipyretic in infants and children, but there is now widespread agreement that use in this age group is contraindicated, except in exceptional circumstances, because of the possible link with the development of Reye's syndrome. It is therefore of concern that a recent review of advice about home management of fever entered on the World Wide Web identified three of 34 documents still recommending aspirin, whilst it was discouraged in only 22 [73]. Its use remains common in some regions, including the countries of the former Soviet Union and Eastern Europe.

Fear and misconceptions about the perceived dangers of fever often lead to unnecessary use of antipyretic drugs. Indeed, fever is generally a protective physiological mechanism, enhancing immune response. Certainly diagnosis and effective therapeutic intervention may be obscured by reduction of raised temperature and relief of symptoms. Aspirin therapy should therefore be reserved for use in adults in whom fever itself may be deleterious, for example because of increased metabolic needs in the malnourished, increased sweating in those who are already dehydrated or increased cardiac demands in those with compromised cardiac function. It may also be justified in those who are likely to experience considerable symptomatic benefits when temperature is lowered, so allowing sleep and improving oral fluid intake. There is, however, only a modest relationship between subjective feelings of improvement and fall in temperature and it is almost never justified to give aspirin to a pyrexial patient who is not in any discomfort. Particular caution is appropriate when fever is likely to have a non-infectious cause, such as lymphoma or other malignancy. There is no difference in the response of fever to aspirin between patients with trivial as opposed to serious infections [72,74].

Comparative studies of the antipyretic effects of aspirin against other agents have mainly been conducted in children rather than adults and are often difficult to interpret because drug dosages have not been comparable or outcome measures have been unsatisfactory [75]. In general, any differences in efficacy between treatment with aspirin and paracetamol or ibuprofen have been small [76,77].

The antipyretic dose of aspirin for adults is 300–600 mg orally every four hours. Rectal administration of suppositories is an alternative if oral medication is a problem. The onset of action is relatively slow and lasts only 3–4 hours [78]. There are now safer (ibuprofen), equally inexpensive (paracetamol) and equally effective alternatives. Although combined use of aspirin with paracetamol will lower temperature to a greater degree than either drug alone, this is of questionable clinical benefit and combination therapy is seldom if ever justified.

It should also be remembered that toxic doses of aspirin in themselves have a pyretic effect, causing sweating and exacerbating dehydration, especially in children and the frail elderly. Treatment should be by external cooling.

RHEUMATIC FEVER

Although this condition is now uncommon, high-dose aspirin has long been used to reduce the fever, inflammatory arthritis and erythrocyte sedimentation rate in patients with rheumatic fever [79] . It may still be the preferred drug in the acute stage, when it appears to be as effective as corticosteroids. Furthermore, a recent meta-analysis failed to demonstrate any advantage of corticosteroid treatment over salicylates in preventing long-term valvular heart disease [80].

BEHÇET'S SYNDROME

Behçet's syndrome is an uncommon multisystem vasculitic disorder characterized by recurrent oral and genital ulcers, uveitis and joint and central nervous system involvement. The use of high-dose aspirin as an anti-inflammatory and analgesic agent has now been superseded by corticosteroids, immunosuppressants and other non-steroidal anti-inflammatory drugs. Patients are also at high risk of arterial and venous thrombotic episodes. Increased platelet aggregation associated with decreased prostacyclin sensitivity of platelets appears to be chiefly responsible [81] and so antiplatelet drugs have a logical role to play in management. Low-dose aspirin has now become a standard part of treatment, both during acute exacerbations and prophylactically [82,83].

KAWASAKI DISEASE

This acute, systemic vasculitic disease of infancy and early childhood is now the commonest cause of acquired heart disease in developed countries. Its aetiology remains unknown, but a bacterial superantigen toxin may be responsible [84]. Also known as mucocutaneous lymph node syndrome, Kawasaki disease is characterized by high fever lasting for at least five days, a diffuse rash, bilateral conjunctivitis, inflamed mucous membranes, erythema and subsequent desquamation of palms and soles and cervical lymphadenopathy. Cardiac involvement may lead to coronary artery aneurysms, predisposing patients to myocardial infarction, dysrhythmias and sudden death [85]. In Britain the diagnosis is often missed and treatment is often delayed or inadequate [86].

Kawasaki disease is one of the few generally accepted indications for use of aspirin in children, though its benefit has not been proven in prospective randomized controlled trials [87]. There is also some uncertainty about the optimum dose and duration, with recommended initial doses ranging from 30 up to 180 mg/kg/day in three or four divided doses. There is widespread agreement that in the acute stage it should always be given together with intravenous immunoglobulin. After 14 days, or once fever has subsided, aspirin can be reduced to an antiplatelet dose of 3–5 mg/kg/day and if no coronary artery abnormalities are detected on echocardiogram after 6–8 weeks, it can then be stopped. If coronary artery lesions are detected at any stage, low-dose aspirin should be continued long term. This regimen leads to more rapid recovery and reduces the risk of potentially life-threatening sequelae [88].

Conclusions

Aspirin remains a highly effective analgesic, anti-inflammatory and antipyretic drug, but its use has largely been overtaken by other non-steroidal agents which may be better tolerated and are often more convenient. It continues to play an important role in management of uncommon but serious inflammatory conditions, such as juvenile arthritis and Kawasaki disease, and is still widely used as a simple analgesic.

KEYPOINTS

- Aspirin provides symptomatic relief of pain and stiffness in inflammatory arthritis and soft tissue injuries. Relatively high doses are necessary for anti-inflammatory effect and these may be associated with symptoms of toxicity.
- As a simple analgesic, aspirin is active against mild to moderate pain of headache, dysmenorrhoea, toothache, postoperative discomfort, mild viral illness and musculoskeletal injury. It is less effective in treating superficial, cutaneous pain and is ineffective in deep, visceral pain.
- Aspirin is an effective antipyretic agent, but its use should be reserved for adults in whom fever itself may be deleterious, such as the malnourished, the dehydrated or those with compromised cardiac function and those in whom the symptomatic benefit can be justified. It should generally no longer be given to children because of the risk of developing Reye's syndrome.

REFERENCES

1. Stoehr GP, Ganguli M, Seaberg EC, Echement DA, Belle S. Over-the-counter medication use in an older rural community: the MOVIES Project. J Am Geriatr Soc 1997; 45: 158–165.
2. Cartwright A, Smith C. Elderly People, Their Medicines and Their Doctors. London: Routledge, 1988.
3. Hale WE, May FE, Marks RG, Stewart RB. Drug use in an ambulatory elderly population: a five year update. Drug Intell Clin Pharmacol 1987; 21: 530–535.
4. Heller CA, Ingelfinger JA, Goldman P. Nonsteroidal anti-inflammatory drugs and aspirin – analysing the scores. Pharmacotherapy 1985; 5: 30–38.
5. Levy G. Clinical pharmacokinetics of aspirin. Pediatrics 1978; 62 (suppl 2): 867–872.
6. Day RO. Aspirin and salicylates. In Kelley WN, Harris ED, Ruddy S (eds) Textbook of Rheumatology, 4th edn. Philadelphia: WB Saunders, 1993: 681–691.
7. Levy G, Giacomini KM. Rational aspirin dosage regimens. Clin Pharmacol Ther 1978; 23: 247–252.
8. Levy G. Clinical pharmacokinetics of salicylates: a reassessment. Br J Clin Pharmacol 1980; 10 (suppl 2) 285S–290S.
9. Makela AL, Yrjana T, Mattila M. Dosage of salicylates for children with juvenile rheumatoid arthritis. Acta Paediatr Scand 1979; 68: 423–430.
10. Bensen WG, Laskin CA, Paton TW, Little HA, Fam AG. Twice-daily dosing of enteric coated aspirin in patients with rheumatic diseases. J Rheumatol 1979; 6: 351–356.
11. Fremont-Smith K, Bayles T. Salicylate therapy in rheumatoid arthritis. JAMA 1965; 192: 1133–1139.
12. Boardman P, Hart F. Clinical measurement of the anti-inflammatory effect of salicylates in rheumatoid arthritis. Br Med J 1967; 4: 264–268.
13. Hadler NM. The argument for aspirin as the NSAID of choice in the management of rheumatoid arthritis. Drug Intell Clin Pharmacol 1984; 18: 34–38.
14. Lee P, Andersen JA, Miller J, Webb J, Buchanan WW. Evaluation of analgesic action and efficacy of antirheumatic drugs. Study of 10 drugs in 684 patients with rheumatoid arthritis. J Rheumatol 1976; 3: 283–294.
15. Gall EP. The safety of treating rheumatoid arthritis with aspirin. JAMA 1982; 247: 63–64.
16. Luggen ME, Gartside PS, Hess EV. Nonsteroidal antiinflammatory drugs in rheumatoid arthritis: duration of use as a measure of relative value. J Rheumatol 1989; 16: 1565–1569.
17. Fries JF, Ramey DR, Singh G *et al.* A reevaluation of aspirin therapy in rheumatoid arthritis. Arch Intern Med 1993; 153: 2465–2471.
18. Kvien TK, Olsson B, Hoyerall HM. Acetylsalicylic acid and juvenile rheumatoid arthritis. Acta Paediatr Scand 1985; 74: 755–759.

19. Baum J. Aspirin in the treatment of juvenile arthritis. Am J Med 1983; 74: 10–15.
20. Doughty RA, Giesecke L, Athreya B. Prospective study of salicylate therapy in juvenile rheumatoid arthritis: dosage, serum salicylate levels, and clinical/biochemical toxicity. Clin Pharmacol Ther 1979; 25: 221–230.
21. Bardare M, Cislaghi GU, Mandeli M, Serini F. Value of monitoring plasma salicylate levels in treating juvenile rheumatoid arthritis. Arch Dis Child 1978; 53: 381–385.
22. Melton JW, Lussier A, Ward JR, Neistadt D, Multz C. Naproxen vs aspirin in osteoarthritis of the hip and knee. J Rheumatol 1978; 5: 338–346
23. Essigman WK, Chamberlain MA, Wright V. Diflunisal in oesteoarthrosis of the hip and knee. Ann Rheum Dis 1979; 38: 148–151.
24. Brandt KD. Effects of non-steroidal antiinflammatory drugs on chondrocyte metabolism in vitro and in vivo. Am J Med 1987; 83: 29–34.
25. Rashad S, Revell P, Hemingway A *et al.* Effect of non-steroidal anti-inflammatory drugs on the course of osteoarthritis. Lancet 1989; 2: 519–522.
26. Slowman-Kovacs S, Albrecht ME, Brandt KD. Effects of salicylate on chondrocytes from osteoarthritic and contralateral knees of dogs with unilateral anterior cruciate ligament transection. Arthritis Rheum 1989; 32: 486–490.
27. Karch J, Kimberly RP, Stahl NI, Plotz PH, Decker JL. Comparative effects of aspirin and ibuprofen in the management of systemic lupus erythematosus. Arthritis Rheum 1980; 23: 1401–1404.
28. Roubey RAS, Hoffman M. From antiphospholipid syndrome to antibody-mediated thrombosis. Lancet 1997; 350: 1491–1493.
29. Godfrey R, Calabro J, Mills D, Maltz BA. A double-blind crossover trial of aspirin, indomethacin and phenylbutazone in ankylosing spondylitis. Arthritis Rheum 1972; 15: 110.
30. Yu TF, Gutman AB. Study of the paradoxical effects of salicylate in low, intermediate and high dosage on the renal mechanisms for excretion of urate in man. J Clin Invest 1959; 38: 1298–1308.
31. Gill JR Jr, Frolich JC, Bowden RE *et al.* Bartter's syndrome: a disorder characterised by high urinary prostaglandins and a dependence of hyperreninemia on prostaglandin synthesis. Am J Med 1976; 61: 43–51.
32. Norby L, Flamenbaum W, Lentz R, Ramwell P. Prostaglandins and aspirin therapy in Bartter's syndrome. Lancet 1976; ii: 604–606.
33. Littlewood JM, Lee MR, Meadow SR. Treatment of Bartter's syndrome in early childhood with prostaglandin synthetase inhibitors. Arch Dis Child 1978; 53; 43–48.
34. WHO. Cancer pain relief. Geneva: WHO, 1986.
35. Schild HO. Applied Pharmacology: Analgesic Drugs. Edinburgh: Churchill Livingstone, 1980: 278–294.
36. British National Formulary. London: British Medical Association and Royal Pharmaceutical Society of Great Britain, 1997: 192.

37. Van Graffenried B, Nuesch E, Maeglin B, Hagler W, Kuhn M. Assessment of analgesics in dental surgery outpatients. Eur J Clin Pharmacol 1980; 18: 479–482.
38. Seymour RA, Rawlins MD. The efficacy and pharmacokinetics of aspirin in postoperative dental pain. Br J Clin Pharmacol 1982; 13: 807–810.
39. Murray WJ. Evaluation of acetaminophen–salicylamide combinations in treatment of headache. J Clin Pharmacol 1967; 7: 150–155.
40. Cooper SA. Comparative analgesic efficacies of aspirin and acetaminophen. Arch Intern Med 1981; 141: 282–285.
41. Mehlisch DR. Review of the comparative analgesic efficacy of salicylates, acetaminophen and pyrazolones. Am J Med 1983; 75: 47–52.
42. Cooper SA, Beaver WT. A model to evaluate mild analgesics in oral surgery. Clin Pharmacol Ther 1976; 20: 241–250.
43. Murray WJ. Evaluation of aspirin in treatment of headache. Clin Pharmacol Ther 1964; 5; 21–25.
44. Parkhouse J, Rees-Lewis M, Skolinik M, Peters H. The clinical dose response to aspirin. Br J Anaesth 1968; 40: 433–441.
45. Seki T. Evaluation of effect of acetylsalicylic acid using electrical stimulation on the forefinger of healthy volunteers. Br J Clin Pharmacol 1978; 6: 521–524.
46. Gruber CM, Bauer RO, Bettigole JR, Lash AF, McDonald JS. A multicenter study for analgesia involving fenoprofen, propoxyphene (alone or in combination) with placebo and aspirin controls in postpartum pain. J Med 1979; 10: 65–97.
47. Cass LJ, Frederick WS. Clinical comparison of a sustained release aspirin. Curr Ther Res 1965; 7: 673–682.
48. Levy G. Aspirin absorption rate and analgesic effect. Anesth Analg 1965; 44: 637–841.
49. Cass LJ, Frederick WS. A clinical evaluation of a sustained release aspirin. Curr Ther Res 1965; 7: 683–692.
50. Tfelt-Hansen P, Olesen J. Effervescent metoclopramide and aspirin (Migravess) versus effervescent aspirin or placebo for migraine attacks: a double-blind study. Cephalalgia 1984; 4: 107–111.
51. The Oral Sumatriptan and Aspirin plus Metoclopramide Comparative Study Group. A study to compare oral sumatriptan with oral aspirin plus oral metoclopramide in the acute treatment of headache. Eur Neurol 1992; 32: 177–184.
52. Hanington E, Jones RJ, Arness JAL, Wachowicz B. Migraine: a platelet disorder. Lancet 1981; ii: 720–723.
53. D'Andrea G, Cananzi A, Toldo M, Cortelazzo S, Ferro-Milone F. Platelet behaviour in classic migraine: responsiveness to small doses of aspirin. Thromb Haemostas 1983; 49: 153.
54. Masel BE, Chesson AL, Peters BH, Levin HS, Alperin JB. Platelet antagonists in migraine prophylaxis. A clinical trial using aspirin and dipyridamole. Headache 1980; 20; 13–18.

55. Grotemeyer KH, Scharafinski HW, Schlake HP, Husted IW. Acetylsalicylic acid vs metoprolol in migraine prophylaxis – a double blind crossover study. Headache 1990; 30: 639–641.
56. Buring JE, Peto R, Hennekens CH. Low-dose aspirin for migraine prophylaxis. JAMA 1990; 264: 1711–1713.
57. Peto R, Gray R, Collins R *et al.* Randomised trial of prophylactic daily aspirin in British male doctors. Br Med J 1988; 296: 313–316.
58. Hugues FC, Lacoste JP, Danchot J, Joire JE. Repeated doses of combined oral lysine acetylsalicylate and metoclopramide in the acute treatment of migrane. Headache 1997; 37: 452–454.
59. Volans GN. The absorption of aspirin during migraine. Br Med J 1974; 4: 265–269.
60. Volans GN. The effect of metoclopramide on the absorption of effervescent aspirin in migraine. Br J Clin Pharmacol 1975; 2: 57–63.
61. Headache Classification Committee of the International Headache Society. Classification and diagnostic criteria for headache disorders, cranial neuralgias and facial pain. Cephalagia 1988; 8 (suppl 7): 1–96.
62. Mathew N, Kurman R, Perez F. Drug induced refractory headache. Headache 1990; 30: 634–638.
63. Schoenen J, De Noordhout AM. Headache. In: Wall PD, Melzack R (eds) Textbook of Pain, 3rd edn. Edinburgh: Churchill Livingstone, 1994: 495–521.
64. Chan WY, Hill JC. Determination of menstrual prostaglandin levels in nondysmenorrhoeic and dysmenorrhoeic subjects. Prostaglandins 1978; 15: 365–375.
65. Klein JR, Litt IF, Rosenberg A, Udall L. The effect of aspirin on dysmenorrhea in adolescents. J Pediatr 1981; 98: 987–990.
66. Rosenwaks Z, Jones GS, Henzl MR *et al.* Naproxen sodium, aspirin, and placebo in primary dysmenorrhea. Reduction of pain and blood levels of prostaglandin F2-alpha metabolite. Am J Obstet Gynecol 1981; 140: 592–598.
67. Marx JL. Dysmenorrhoea: basic research leads to a rational therapy. Science 1979; 205: 175–176.
68. King RB. Concerning the management of pain associated with herpes zoster and of postherpetic neuralgia. Pain 1988; 33: 73–78.
69. De Benedittis G, Besana F, Lorenzetti A. A new topical treatment for acute herpetic and postherpetic neuralgia: the aspirin/diethyl ether mixture. An open-label study plus a double blind controlled clinical trial. Pain 1992; 48: 383–390.
70. De Benedittis G, Lorenzetti A. Topical aspirin/diethyl ether mixture versus indomethacin and diclofenac/diethyl ether mixtures for acute herpetic neuralgia and postherpetic neuralgia: a double-blind crossover placebo-controlled study. Pain 1996; 65: 45–51.
71. Tharion G, Bhattacharji S. Aspirin in chloroform as an effective adjuvant in the management of chronic neurogenic pain. Arch Phys Med Rehab 1997; 78: 437–439.

72. Walson PD. Fever. In: Rakel RE (ed.) Conn's Current Therapy. Philadelphia: WB Saunders, 1997: 24–27.
73. Impicciatore P, Pandolfini C, Casella N, Bonati M. Reliability of health information for the public on the world wide web: systematic survey of advice on managing fever in children at home. Br Med J 1997; 314: 1875–1881.
74. Rosenthal TC, Silverstein DA. Fever. What to do and what not to do. Postgrad Med 1988; 83: 75–84.
75. Temple AR. Review of comparative antipyretic activity in children. Am J Med 1983; 75: 38–46.
76. Aksoylar S, Aksit S, Caglayan S *et al.* Evaluation of sponging and antipyretic medication to reduce body temperature in febrile children. Acta Paediatr Jpn 1997; 39: 215–217.
77. Autret E, Reboul-Marty J, Henry-Launois B *et al.* Evaluation of ibuprofen versus aspirin and paracetamol on efficacy and comfort in children with fever. Eur J Clin Pharmacol 1997; 51: 367–371.
78. Vigano A, Dalla Villa A, Cecchini J, Biasini GC, Principi N. Correlation between dosage and antipyretic effect of aspirin in children. Eur J Pharmacol 1986; 31: 359–361.
79. Bywaters E, Thomas G. Bedrest, salicylates, and steroids in rheumatic fever. Br Med J 1961; 524: 1628–1634.
80. Albert DA, Harel L, Karrison T. The treatment of rheumatic carditis: a review and meta-analysis. Medicine (Baltimore) 1995; 74: 1–12.
81. Wilson AP, Efthimiou J, Betteridge DJ. Decreased prostacyclin sensitivity of platelets in patients with Behçet's syndrome. Eur J Clin Invest 1988; 18: 410–414.
82. Huong DL, Dolmazon C, De Zuttere D *et al.* Complete recovery of right intraventricular thrombus and pulmonary arteritis in Behçet's disease. Br J Rheum 1997; 6: 130–132.
83. Rewald E, Jaksic JC. Behçet's syndrome treated with high-dose intravenous IgG and low-dose aspirin. J Roy Soc Med 1990; 83: 652–653.
84. Curtis N. Kawasaki disease. Br Med J 1997; 315: 322–323.
85. Kawasaki T. Acute febrile mucocutaneous lymph node syndrome with lymphoid involvement with specific desquamation of the fingers and toes in children. Jpn J Allergy 1967; 16: 178–222.
86. Dhillon R, Newton L, Rudd PT, Hall SM. Management of Kawasaki disease in the British Isles. Arch Dis Child 1993; 69: 631–636.
87. Newburger JW. Treatment of Kawasaki disease. Lancet 1996; 347: 1128.
88. Nakashima L, Edwards DL. Treatment of Kawasaki disease. Clin Pharmacol 1990; 9: 755–762.

Therapeutic use in vascular diseases

Introduction

The past 10–15 years have seen an enormous growth in the use of low-dose aspirin prophylaxis for the prevention of cardiovascular disease [1–3]. In high-risk subjects, aspirin reduces cardiovascular events by about one-third or about 40 events per 1000 patients treated per year [2]. However, its use is associated with important adverse events and the balance of risks and benefits varies in different clinical situations, especially when taken as a primary preventive measure.

EPIDEMIOLOGICAL STUDIES

Initial interest in the potential beneficial effects of aspirin for lowering the risk of cardiovascular disease arose from the findings of several observational epidemiological studies conducted in the mid-1970s. The most important was a case control study conducted by the Boston Collaborative Drug Surveillance Group in survivors of myocardial infarction [4]. This found reduced risk of recent myocardial infarction or death associated with regular aspirin intake, as did a smaller study involving only men [5]. In contrast, two other studies found no association between aspirin use and fatal coronary disease [6] or risk of myocardial infarction [7] and a later study [8] even suggested that daily use of aspirin by elderly people increased the risk of developing ischaemic heart disease. However, it seems far more likely that the association was non-causal and may merely reflect the greater likelihood of at-risk patients having started to take preventive measures.

Primary prevention of occlusive cardiovascular disease

HEALTHY (LOW-RISK) SUBJECTS

The potential benefits of low-dose aspirin therapy for the primary prevention of occlusive cardiovascular disease in asymptomatic subjects considered to be at low risk have been studied in two large randomized controlled trials, one in Britain [9] and one in America [10]. Together they included about 28 000 healthy, middle-aged and older subjects, all of whom were male. Whilst there is at present no evidence from randomized controlled trials in healthy women, a prospective cohort study of 87 000 registered nurses has examined the association between self-selected aspirin use and risk of a first myocardial infarct [11]. This suggested potential benefit in those taking between one and six aspirins a week, especially if aged 50 years or older. The apparent benefits of moderate aspirin usage were accentuated in current smokers and those with a history of hypertension or elevated serum cholesterol. A randomized controlled trial in 40 000 women health professionals aged over 45 years and with no previous history of cardiovascular disease (the Women's Health Study) is ongoing [12]. This should provide more reliable data concerning the balance of benefits and risks in this population.

The completed randomized controlled trials in men [9,10] came to slightly different conclusions. This probably reflected their inadequate statistical power to yield conclusive results, given the very low rate of cardiovascular events in the particularly healthy subjects studied, all of whom were themselves physicians. The cardiovascular mortality rate was in fact only 15% of that expected for the general population of matched age and sex over a similar period. Less than 1% of the subjects had a cardiovascular event each year and so the potential for a reduction in absolute risk was very small [13].

In the British Doctors' Trial [9], 5139 subjects aged from 50 to 78 years were randomized in a ratio of 2:1 to take aspirin 500 mg daily or to avoid aspirin-containing products. They were followed up for six years, by which time there was only a non-significant 12% reduction in incidence of myocardial infarction and no statistically significant difference in the combined endpoint of important cardiovascular events, myocardial infarction, stroke and total cardiovascular mortality. The non-significant 10% reduction in total mortality was mainly due to a non-significant 15% reduction in non-vascular deaths. There was a slight increase in the numbers of disabling strokes in the aspirin-treated subjects, outweighing a lower incidence of transient ischaemic attacks.

In the American Physicians' Health Study [10], 22 071 subjects aged between 40 and 84 years were given either aspirin 325 mg on alternate days or placebo. Few subjects were aged over 75 years. The trial was stopped early, after five years of follow-up, because of a significant reduction in the incidence of fatal and non-fatal myocardial infarction. This was reduced by 44% in the aspirin group, but the reduction was only apparent in the over-50s and there was no overall reduction in cardiovascular deaths, which was the primary objective of the study. A subsequent subgroup analysis showed a significant 51% reduction in myocardial infarction in patients who adhered closely to the trial protocol and took their aspirin regularly, but a non-significant 17% reduction in those who took less than half of their study tablets. The reduction in myocardial infarction was most significant during the early morning [14], when platelet aggregability and so aspirin sensitivity is greatest [15]. As in the British study, there was a non-significant increase in haemorrhagic stroke and sudden death in the aspirin-treated group, which offset the reduction in fatal myocardial infarction.

The absolute risk of cardiovascular events in healthy subjects is very low and so adverse effects of any preventive treatment must be negligible if the net effect is to be beneficial. The apparent excess of cerebral haemorrhage in both published trials is therefore of particular importance and there is considerable debate about whether the possible small beneficial effects of long-term low-dose aspirin therapy (about one important cardiovascular event prevented per year for every 1000 patients treated) outweigh the hazards when used for primary prevention in asymptomatic subjects [16] (Figure 5.1). Despite the flimsy evidence in support of its use, increasing numbers of health-conscious, physically active middle-aged and older men and women are starting to take regular aspirin as prophylaxis against cardiovascular disease [17]. Evidence that it may also have a role in primary prevention of cataract, certain cancers and cognitive decline may lend further support to this trend.

At present, no major authority recommends routine universal aspirin prophylaxis [18–20]. However, new guidelines from the Australian National Health and Medical Research Council advise doctors to consider low-dose aspirin therapy for all patients over 50 years and many other organizations have their recommendations under review. Whether asymptomatic patients will comply with regularly taking aspirin over many years is uncertain.

ASYMPTOMATIC SUBJECTS AT INCREASED RISK OF CARDIOVASCULAR DISEASE

The role of aspirin therapy in asymptomatic patients who have multiple vascular risk factors (e.g. smoking, hypercholesterolaemia and hypertension) remains unclear. There is no good evidence from randomized controlled

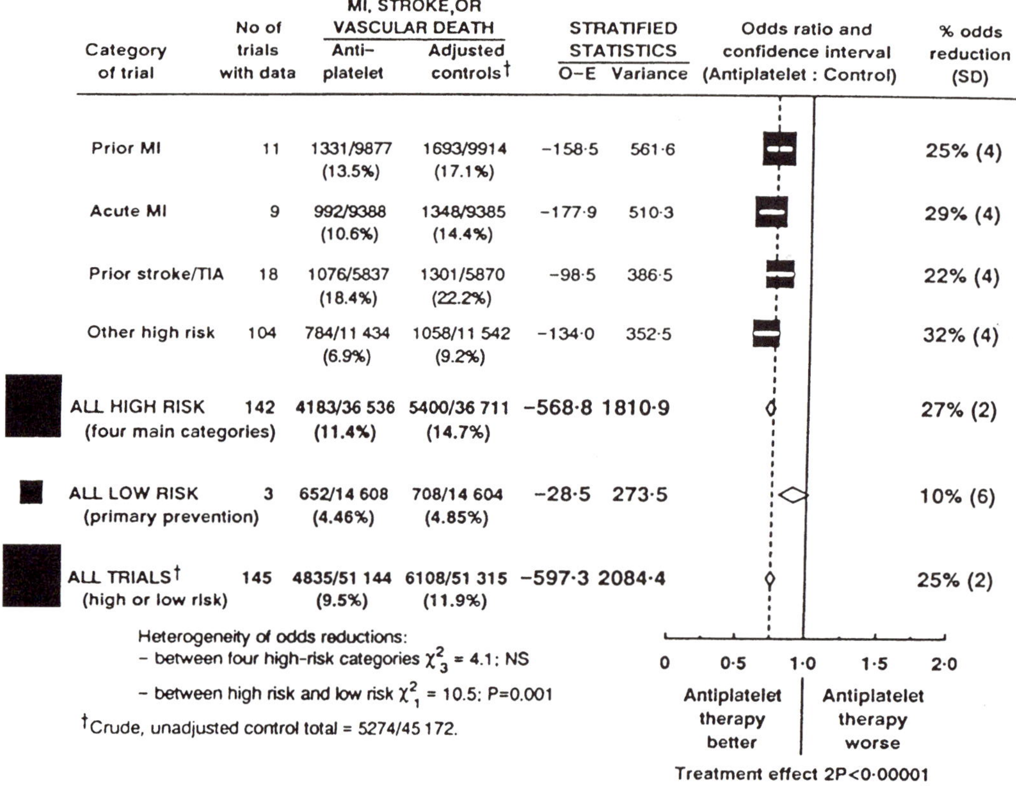

Category of trial	No of trials with data	MI, STROKE, OR VASCULAR DEATH: Anti-platelet	MI, STROKE, OR VASCULAR DEATH: Adjusted controls†	STRATIFIED STATISTICS: O–E	STRATIFIED STATISTICS: Variance	% odds reduction (SD)
Prior MI	11	1331/9877 (13.5%)	1693/9914 (17.1%)	−158·5	561·6	25% (4)
Acute MI	9	992/9388 (10.6%)	1348/9385 (14.4%)	−177·9	510·3	29% (4)
Prior stroke/TIA	18	1076/5837 (18.4%)	1301/5870 (22.2%)	−98·5	386·5	22% (4)
Other high risk	104	784/11 434 (6.9%)	1058/11 542 (9.2%)	−134·0	352·5	32% (4)
ALL HIGH RISK (four main categories)	142	4183/36 536 (11.4%)	5400/36 711 (14.7%)	−568·8	1810·9	27% (2)
ALL LOW RISK (primary prevention)	3	652/14 608 (4.46%)	708/14 604 (4.85%)	−28·5	273·5	10% (6)
ALL TRIALS† (high or low risk)	145	4835/51 144 (9.5%)	6108/51 315 (11.9%)	−597·3	2084·4	25% (2)

Fig. 5.1 Proportional effects of antiplatelet therapy on vascular events (myocardial infarction, stroke or vascular death) in four main high-risk categories of trial and in low risk (primary prevention). O − E = observed minus expected (from reference [2] with permission).

trials to guide management, but it would seem that the protective effects of aspirin are of similar magnitude whether risk factors are present or not [10]. Thus, as the event rate inevitably will be higher in patients with risk factors, the absolute benefit of aspirin treatment is also likely to be greater. Several relevant studies are in progress. The Thrombosis Prevention Trial is studying low-dose aspirin, with or without low-dose warfarin, versus placebo in 7500 middle-aged men at high risk of ischaemic heart disease [21,22] and the Hypertension Optimal Treatment Study is investigating the role of low-dose aspirin in 18 000 hypertensive patients [23].

At present the focus of preventive effort in asymptomatic patients should be on modifying the primary risk factors for cardiovascular disease. In those who lack contraindications to aspirin use (e.g. allergy, peptic ulceration, risk factors for bleeding, etc.) it is possible that the additional benefits of prophylactic low-dose aspirin may outweigh any disadvantage. However, the decision to treat should be always be made on an individual basis after a full consideration of the potential benefits and risks [17,19].

PATIENTS WITH CHRONIC STABLE ANGINA

The use of low-dose aspirin is now accepted as standard treatment for the great majority of patients with chronic stable angina [24], although chronic ischaemic heart disease without previous infarction is generally not associated with platelet hypersensitivity [25]. The Antiplatelet Trialists' Collaboration [2] identified only four randomized controlled trials of aspirin involving about 2500 patients with this indication. By far the largest of these was the Swedish Angina Pectoris Aspirin Trial [26] in which 2035 patients with chronic stable angina but without previous myocardial infarction were randomized to aspirin 75 mg daily or placebo for 50 months. There was a highly significant reduction of 34% in myocardial infarction and sudden death. Reduction in other oucome events (vascular death, all-cause mortality, vascular events, stroke) ranged from 22% to 32%. Thus treating 1000 angina patients with aspirin may prevent about 13 important cardiovascular events a year.

A subgroup analysis of 333 men with chronic stable angina included in the Physicians' Health Study suggested that the reduction in risk of myocardial infarction associated with aspirin therapy was at least as great as in the asymptomatic trial subjects, although the frequency and severity of anginal attacks were unchanged [27]. Moreover, aspirin did not influence the incidence or delay onset of new cases of confirmed angina [28].

Results from the British Regional Heart Study collected in 1992 reported that overall, only 29% of middle-aged men with angina were taking regular prophylactic aspirin, although use was much more likely in those who had undergone coronary artery bypass surgery or angioplasty or had suffered recent symptoms [29]. More recent audits of ischaemic heart disease patients in both primary [30,31] and secondary care [32] have also suggested that the situation is improving, whilst confirming that aspirin is still underused. Simple interventions such as feedback to general practitioners of their prophylactic aspirin prescribing can significantly increase the proportion of patients appropriately taking medication [33].

PATIENTS WITH PERIPHERAL VASCULAR DISEASE

The Antiplatelet Trialists identified 27 trials in patients with intermittent claudication, but only five of these tested aspirin, all using high doses (975 mg daily or more) [2]. The overall benefit of therapy appeared to be similar to that seen with other patient groups at risk but because patient numbers were small, the risk reduction was not statistically significant. An analysis of results from the US Physicians' Health Study found that long-term aspirin therapy reduced need for peripheral arterial surgery [34], but there is little evidence that aspirin has a significant symptomatic effect on claudication distance or severity of pain in patients with peripheral vascular disease and in those requiring surgery, perioperative aspirin in a dose of 75 mg daily failed to inhibit platelet hyperactivity, as measured by *ex vivo* shear-induced haemostasis and collagen-induced thrombus formation [35]. Nevertheless, it is generally accepted that it is appropriate to consider prophylactic low-dose aspirin in all patients with evidence of peripheral arterial disease, to reduce their increased risk of acute vascular events [36].

PATIENTS WITH ATRIAL FIBRILLATION

Patients with atrial fibrillation are at significantly increased risk of thromboembolic complications, most notably ischaemic stroke. The Framingham Study estimated that overall risk was increased fivefold, with an annual incidence of stroke of about 5% [37]. Those at greatest risk of thromboembolic events are the elderly, those with recent congestive cardiac failure or left ventricular dysfunction on echocardiography, those with previous history or evidence of thromboembolism and those who are also hypertensive [38,39].

Patients with atrial fibrillation associated with valve disease have long been treated with long-term oral anticoagulants, occasionally combined with aspirin in very high-risk patients [40]. The treatment of non-rheumatic atrial fibrillation has recently attracted much interest and the role of aspirin has been considered in four major randomized controlled trials [41–44] and in a predetermined subgroup analysis of a fifth [45]. It has also been the subject of numerous reviews [46–48].

The Copenhagen Atrial Fibrillation, Aspirin, Anticoagulation Study (AFASAK) [41] compared aspirin 75 mg daily, warfarin (INR 2.8 and 4.2) and placebo in 1007 patients over two years. It was the first prospective study to be published in 1989. There was a non-significant risk reduction in primary thromboembolic events of 14% in the aspirin group and a significant reduction of 59% in the warfarin-treated patients. However, whilst major bleeding complications were a problem in the warfarin patients (3.1% per year), they were only slightly more common in the aspirin patients (0.3% per year) than in those on placebo (0.0% per year).

The SPAF I study [42] compared aspirin 325 mg daily or placebo in warfarin-ineligible patients. Active treatment reduced the relative risk of ischaemic stroke significantly by 42%, from an absolute annual rate of 6.3% in placebo patients to 3.6% in aspirin-treated patients. The risk of major bleeding did not significantly differ between groups. The subsequent SPAF II study [43] directly compared aspirin 325 mg daily and warfarin (INR between 2.0 and 4.5). Aspirin appeared to be at least as effective as warfarin in reducing overall incidence of disabling strokes (2.5% per year versus 2.2%), though there were overall more cerebral embolic events in the aspirin group (2.7% per year versus 1.9%). However, neither of these differences was statistically significant. In those aged less than 75 years, warfarin was associated with an event rate of 1.3% per year, compared with 1.9% for aspirin-treated patients. In those older than 75 years, the event rates were 3.6% and 4.8% respectively, though warfarin was associated with a significantly greater risk of major haemorrhage (4.2% versus 1.6%), including rates of intracranial bleeding (1.8% and 0.8%). Interpretation of the results of this study is difficult because of the relatively high target range for the INR and the lack of a placebo group. Nevertheless, in lower risk patients, only one in 200 aspirin-treated patients suffered a stroke or other embolic event each year.

The Boston Area Anticoagulation Trial for Atrial Fibrillation [45] randomized patients to low-intensity warfarin (INR between 1.5 and 2.7) or no treatment, with a mean follow-up of 2.2 years. Nearly half of the control group were on aspirin, most of whom were taking 325 mg daily. A subgroup analysis comparing controls on aspirin with controls not taking aspirin showed a non-significant increase in risk of ischaemic stroke of 89%. The annual stroke rate in aspirin control patients was 3.9% versus 1.8% in non-aspirin controls.

The European Atrial Fibrillation Trial (EAFT) was a secondary prevention trial in patients who had already suffered a cerebrovascular event [49]. Among patients assigned to aspirin 300 mg daily, the annual incidence of an acute vascular event was 15%, against 19% in those on placebo. However, in the 40% of patients without contraindications to anticoagulation, warfarin therapy was significantly more effective than aspirin, with only an 8% annual rate of outcome events. The incidence of major bleeding events was 0.9% per year on aspirin and 2.8% per year on anticoagulants.

Aspirin alone does not significantly reduce laboratory indices of haemostasis and clotting in patients with atrial fibrillation [50] and theoretically it was suggested that combination therapy of aspirin and low-intensity warfarin might be most advantageous; the aspirin would reduce platelet emboli from ulcerated plaques, whilst warfarin would be more effective against emboli from fibrin-rich thrombi in dilated left atria or poorly contractile left ventricles. Unfortunately, the recent SPAF III Study [44] showed that aspirin (325 mg daily) was certainly not able to compensate for reduced anticoagulation (INR less than 1.5) associated with low-intensity warfarin regimens. This study compared the effectiveness of aspirin and low-intensity, fixed-dose warfarin with conventional adjusted-dose warfarin in high-risk patients with atrial fibrillation. The study was stopped early because of an increased risk of 301% of ischaemic stroke or systemic embolism in the aspirin plus low-dose warfarin patients compared with those receiving adjusted-dose warfarin.

In summary, present evidence suggests that prophylactic use of low-dose aspirin may be appropriate for low-risk patients with non-rheumatic atrial fibrillation (e.g. normotensives aged under 75 years, with normal left ventricular function and no previous thromboembolic events or history of diabetes [46]) as well as for those in whom full anticoagulation is contraindicated, but aspirin is not [51]. It is likely to be more efficacious than placebo, with an analysis of pooled data from randomized controlled trials suggesting an overall relative risk reduction of stroke of about 21%, although the 95% confidence intervals were from 0% to 38% [48]. It has the advantages of relative safety, low cost and ease of administration. However, aspirin is associated with at least double the rate of thromboembolic events as occurs with warfarin therapy. In higher risk patients, therefore, anticoagulation rather than aspirin should always be the first choice of treatment, despite the significantly increased risk of bleeding. If the risk of haemorrhage is high, then aspirin offers a safer but less effective alternative to warfarin [52].

PATIENTS WITH DIABETES MELLITUS

Patients with both type I (insulin-dependent) and type II (non-insulin dependent) diabetes mellitus are at significantly increased risk of microvascular renal and retinal problems and macrovascular complications, including myocardial infarction and ischaemic stroke [53]. In part, this may be explained by an enhanced tendency to thrombosis due to increased platelet activity and production of thromboxane, which may be effectively blocked by very low-dose aspirin [54]. Some *ex vivo* studies, however, have suggested that platelets from patients with insulin-dependent diabetes may be resistant to the effects of aspirin by mechanisms independent of the COX pathway of aggregation and so antiplatelet treatment with aspirin to prevent thromboembolic events might therefore be less effective in diabetic patients [55].

Platelet hyperactivity has been implicated in the pathogenesis of diabetic retinopathy and, in experimental models, antiplatelet therapy has prevented the development of retinal vascular abnormalities when given from the first day after onset of diabetes [56]. The hyperfiltration which precedes the decline in glomerular filtration rate in long-term diabetes has also been attributed to prostaglandin activity and it has been suggested that aspirin therapy may therefore protect against the development and progression of diabetic nephropathy [57].

Epidemiological studies have suggested some potential protective effects of aspirin usage, although not against the development of diabetic retinopathy [58]. Clinical trial results have been inconsistent, at least in part due to small patient numbers and unrealistic outcome measures. The Veterans Administration Co-operative Study Group compared aspirin 975 mg daily plus dipyridamole and placebo in patients with established gangrene or limb amputation and found no significant difference in terms of subsequent amputation [59] and a possible increase in sudden death [60]. The DAMAD Study reported no difference in vascular deaths between patients taking aspirin 975 mg daily, aspirin plus dipyridamole or placebo [61]. However, a subgroup analysis by the Antiplatelet Trialists' Collaboration of trials up to 1990 [2] found similar reduction in risk of vascular events associated with antiplatelet therapy in diabetic and non-diabetic subjects.

The only substantial published clinical trial examining the potential benefits of prophylactic use of aspirin specifically in patients with diabetes mellitus is the Early Treatment Diabetic Retinopathy Study [62]. This compared a relatively high dose of 650 mg of aspirin daily with placebo in 3711 patients. All were aged under 70 years and half had a history of cardiovascular disease. After an average duration of treatment of five years, aspirin-treated patients had a borderline significant 28% reduction in myocardial infarction, a non-significant 16% increase in risk of stroke and a significant overall 18% reduction in important cardiovascular events. It was concluded that the effects of aspirin were comparable to those seen in non-diabetic patients and supported recommendation of aspirin to all diabetic patients at increased risk of cardiovascular disease.

The results of trials investigating the effects of aspirin therapy on the development of microvascular complications have also been contradictory. The DAMAD Study [61] showed a significant delay in the progression of non-proliferative retinopathy, with only a small increase over three years in the number of microaneurysms in patients on active treatment compared to a much greater increase on placebo. The benefit was evident in both insulin-dependent and non-insulin dependent patients. However, the ETDRS Study [62] showed no difference in progression of retinopathy after five years treatment with either aspirin 650 mg daily or placebo. There is no evidence that aspirin increases the occurrence of vitreous or preretinal haemorrhages in patients with established diabetic retinopathy [63].

Uncontrolled studies have suggested that albuminuria and the progression of diabetic nephropathy may be delayed by aspirin therapy [64,65] but there is no reliable evidence from randomized controlled trials. A recent small randomized controlled trial of aspirin 100 mg daily in patients with non-insulin dependent diabetes mellitus concluded that treatment may reduce glyco-oxidative damage, as evidenced by significantly reduced levels of skin pentosidine after one year of therapy, but showed no differences in clinical status [66].

The optimum role of aspirin in diabetes mellitus is therefore still to be defined and its use in patients without evidence of vascular disease or other risk factors may not be justified [67].

PATIENTS WITH PROSTHETIC HEART VALVES

Thromboembolism is a potential risk following heart valve replacement with all older and newer synthetic and bioprosthetic valves. Aspirin therapy is usually started a few days after operation, together with oral anticoagulants. A Canadian study comparing warfarin alone or with aspirin (100 mg daily) found a significant reduction in major embolic events and no major bleeding complications in patients taking combined therapy (1.9% per year versus 8.5%) [68].

PATIENTS WITH THROMBOCYTOSIS

Low-dose aspirin is increasingly being prescribed to patients with polycythaemia vera, essential thrombocythaemia and thrombocythaemic states associated with myeloproliferative disorders [69]. All are at significantly increased risk of arterial thrombotic disease and microcirculatory disturbances and an increase in thromboxane synthesis and platelet activation which may be suppressed by aspirin has been demonstrated [70,71].

An initial study of aspirin in patients with polycythaemia vera suggested that it was not only lacking in efficacy, but was also associated with an unacceptable excess of gastrointestinal haemorrhagic complications [72]. However, the relatively small number of patients studied meant that this study lacked statistical power to provide a reliable estimate of effect size and the daily dose of 900 mg aspirin was excessive. Other studies using lower aspirin doses have found it to be well tolerated [73], but there remains no conclusive evidence from randomized controlled trials that long-term low-dose aspirin therapy is a useful antithrombotic strategy in these patients. A pilot study of one year's prophylaxis with aspirin 40 mg daily or placebo in 112 patients with polycythaemia vera [74] has confirmed that such treatment is safe and effective in inhibiting platelet COX activity and a larger trial (the European Collaboration on Low-dose Aspirin in Polycythaemia Vera – ECLAP) is being organized [71].

The syndrome of erythromelalgia and the transient neurological and ocular disturbances sometimes associated with myeloproliferative disease are generally very sensitive to aspirin therapy, with complete remission of symptoms for several days after just a single dose [75,76]. Other antiplatelet drugs and anticoagulants are ineffective and as production of thromboxane

A2 appears to be enhanced in these patients, the beneficial effects of aspirin may be attributed to platelet COX inhibition [77].

Prospective clinical evaluation of low-dose aspirin in patients with essential thrombocythaemia has yet to be conducted. It had been suggested [69] that somewhat higher doses of aspirin may be required than in patients with polycythaemia vera, particularly in those with more marked thrombocytosis, although the condition may also be associated with a more pronounced bleeding tendency [78]. However, a retrospective analysis of a cohort of 68 patients suggested that aspirin, particularly in lower doses (100 mg daily), was safe and effective in reducing thrombotic episodes, with an acceptable risk of bleeding, if applied to patients with a platelet count less than 1000 and/or absence of bleeding history [79].

Secondary prevention of occlusive cardiovascular disease

UNSTABLE CORONARY ARTERY DISEASE

Platelet activation is characteristic of unstable angina patients, leading to thrombus proliferation and potential acute vascular occlusion [80,81]. The impact of aspirin therapy on the risk of myocardial infarction or death in patients with unstable angina has been studied in several large randomized trials conducted during the last 15 years. Both during the acute stages [82] and after treatment for up to 18 months [83,84], aspirin significantly improved outcome, with about a 50% reduction in fatal and non-fatal myocardial infarction within three months [2].

Most studies have used probably unnecessarily large doses of aspirin (up to 1300 mg daily) and the Research Group on Instability in Coronary Artery Disease in Southeast Sweden (RISC) study is the only trial to study low-dose therapy (aspirin 75 mg daily) [85,86]. In this study of 796 men under age 70 with unstable angina or non-Q wave myocardial infarction, whilst there was no significant effect in the first 48 hours, the risk of myocardial infarction and/or death after low-dose aspirin treatment for five days was reduced by 57% (and 75% when given with heparin). The risk reduction after three and 12 months was 64% and 48% respectively, which is at least as great as the effects seen using larger doses of aspirin. Furthermore, subgroup analysis of the RISC Study [87] suggested that benefits of aspirin were evident in patients with silent myocardial ischaemia demonstrated during predischarge exercise tests, as well as those with symptomatic angina. The rebound phenomenon which may occur when heparin is discontinued is prevented by aspirin [88]. It seems likely that administration of an initial 'loading' dose of aspirin (150 mg or more) will ensure early onset of antiplatelet activity.

SUSPECTED EVOLVING MYOCARDIAL INFARCTION

The Second International Study of Infarct Survival (ISIS-2) investigated the benefits of giving aspirin acutely in patients hospitalized for suspected evolving myocardial infarct. A total of 17 187 patients were randomized within 24 hours of symptom onset to receive either 162.5 mg of oral aspirin daily for one month or placebo (with or without an initial infusion of intravenous streptokinase) [89] (Figure 5.2). After five weeks, aspirin was associated with a highly significant 49% reduction in non-fatal reinfarction, 46% reduction in non-fatal stroke and 23% reduction in total vascular mortality. This represents the avoidance of about 25 early deaths, 10 non-fatal reinfarctions and three non-fatal strokes for every 1000 patients treated. Benefit was similar in both men and women and there was no significant increase in haemorrhagic stroke or major bleeding and only a small increase in minor bleeding. The benefit of aspirin was additive to that of thrombolytic therapy, with patients receiving both aspirin and streptokinase having a 42% decrease in vascular deaths. Long-term follow-up after several years showed that the early benefits were sustained [90]. A subsequent meta-analysis has further confirmed the benefits of aspirin given

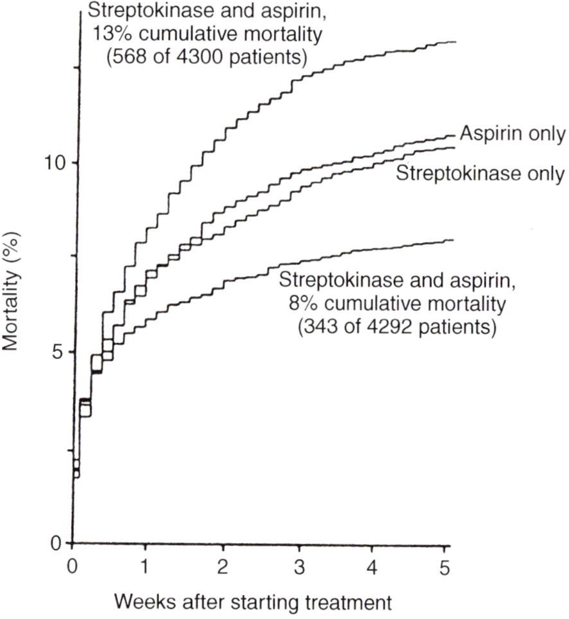

Fig. 5.2
Cumulative mortality from vascular causes up to day 35 in the ISIS-2 Trial of aspirin, streptokinase, both or neither (placebo) in 17 187 patients with suspected acute myocardial infarction (from reference [89] with permission).

with thrombolytic drugs in reducing coronary reocclusion and recurrent ischaemia [91].

Publication of the ISIS-2 results had a major impact on clinical practice, with use of aspirin following acute myocardial infarction increasing from 40% or less of all patients up to 72% in the United States [13] and up to 85% in the United Kingdom [92]. However, as only a very small proportion (1–2%) of patients have clear medical contraindications to its use [92], it should be possible to give aspirin to virtually all patients with acute myocardial infarction, thus avoiding a further several thousand unnecessary premature deaths each year [93]. There is evidence to suggest particular undertreatment of women and elderly patients, who have the highest mortality and so stand to benefit most from early intervention [94].

Acute myocardial infarction with ST segment elevation is nearly always associated with rupture of an atheromatous coronary artery plaque, with release of prothrombotic substances including adenosine diphosphate and collagen. It is assumed that aspirin acts by inhibiting the subsequent platelet aggregation and thrombus formation, although the exact mechanism by which it has its acute effects is unclear [95]. It is known that in patients with very recent infarction, the platelets are especially prone to aggregation, possibly because of an increase in the number of thromboxane receptors [96] and/or a reduced number of prostacyclin receptors [97]. Thrombolytic therapy also increases platelet activation, as demonstrated by a greater than 20-fold increase in thromboxane secretion in urine [98–100], and aspirin can significantly reduce the tendency to reocclusion, reinfarction and death after thrombolysis [89]. It seems to be at least as effective as anticoagulation and considerably safer [101,102].

It is recommended that aspirin therapy should be initiated with a loading dose of at least 150 mg of a soluble formulation [103], which should be chewed rather than swallowed or crushed and given with a drink of warm water. Enteric-coated or controlled-release formulations are not appropriate in this indication. A foil-packed, chewable aspirin preparation is particularly easy to administer [104].

As protection is probably greatest in patients who receive aspirin early after the onset of symptoms [89,105], it should be given immediately, ideally before waiting for hospital admission or final confirmation of diagnosis [106]. If this is not possible, then it is still worth starting aspirin in all patients, however late after onset of symptoms. Unfortunately, various studies have shown significant underuse of aspirin by general practitioners [107,108] and a survey conducted in 1994 reported that only 71% of doctors' bags contained aspirin [109]. A recent American study found that less than half of patients with suspected acute myocardial infarction seen in the emergency departments of four university-affiliated hospitals were given aspirin and in half of those to whom it was given, the delay was greater than an hour [110]. It has therefore been suggested that those at especially high risk of myocardial infarction or stroke should be advised to always carry an aspirin tablet in a jacket pocket or handbag to be taken immediately chest pain or symptoms suggestive of myocardial infarction are experienced [104]. Such an approach would be likely to be cost effective and safe and easily discontinued if an acute coronary event were subsequently ruled out.

SURVIVORS OF MYOCARDIAL INFARCTION

During the 1970s, several trials were organized to investigate the potential of aspirin in the secondary prevention of vascular disease. The first, organized by Peter Elwood and Archie Cochrane from the Medical Research Council Epidemiology Unit in Cardiff, suggested a possible 25% reduction in total mortality at 12 months after hospital admission, but the relatively small patient numbers and event rate meant that conventional statistical significance was not achieved [111]. It was concluded that further trials were urgently required to determine whether or not the effect was real. Subsequent trials used dose regimens varying from 300 to 1500 mg daily and included patients up to five years after their first myocardial infarction [112–117]. Whilst each individual trial again suggested possible benefit, with a reduction in reinfarction rate and/or overall mortality of between 10% and 30%, none of the differences was statistically significant [118]. Only the later PARIS-II Trial, which compared a combination of 975 mg aspirin daily plus dipyridamole against placebo, showed a statistically significant reduction in fatal and non-fatal reinfarction [119].

Together, the trials had recruited many thousands of patients and it required a statistical overview to have adequate statistical power to clearly demonstrate the benefits of long-term prophylaxis. The most recent of these, the Antiplatelet Trialists' Collaborative overview [2], included eight trials of aspirin alone or in combination with dipyridamole, involving about 16 000 patients with a prior history of myocardial infarction (Figure 5.1). Overall there were statistically significant reductions of 31% for non-fatal reinfarction, 39% for non-fatal stroke, 15% for vascular death and 25% for the combined endpoint of important vascular events. Comparable reductions were seen in various patient subgroups, including men and women, middle aged as well as older patients, hypertensives and normotensives and in diabetics and non-diabetics. The optimum duration of treatment was uncertain. In the absence of evidence from direct randomized comparisons of different durations of antiplatelet therapy, it is generally suggested that aspirin therapy should be continued indefinitely, unless some clear contra-indication develops.

In 1992, the Regional Heart Study reported that only 44% of men with previous myocardial infarction were taking aspirin daily [30]. However, a British Cardiac Society survey conducted in 1996 of patients interviewed more than six months after a myocardial infarct reported that 85% were taking aspirin [120]. There is some evidence that women are less likely to be using aspirin after myocardial infarction than men [121].

TRANSIENT ISCHAEMIC ATTACK AND SURVIVORS OF STROKE

Patients with a previous history of stroke or transient ischaemic attack (TIA) have a 13–15-fold excess risk of stroke compared to a similar age- and sex-matched population with no history of stroke [122,123]. They are also at high risk of cardiac events.

Early trials of aspirin after transient ischaemic attack used high doses of aspirin (e.g. 1200–1300 mg daily) and recruited relatively small numbers of patients [124,125]. Despite their equivocal results, they were considered sufficiently encouraging for larger multicentre studies to be organized. The European Stroke Prevention Study [126] compared 325 mg aspirin plus dipyridamole 75 mg daily with placebo in 2500 carefully selected patients with a recent acute cerebral event. After two years treatment, there was a significant reduction in both recurrent stroke and deaths from other causes. The Swedish Aspirin Low-dose Trial (SALT) used a daily dose of 75 mg aspirin and reported a 16–18% reduction relative to placebo in non-fatal or fatal stroke, transient ischaemic attack or myocardial infarction. There was also an excess of gastrointestinal and intracerebral haemorrhage in the active treatment group [127].

The Antiplatelet Trialists' Collaboration was able to include 21 trials of antiplatelet drugs after stroke or transient ischaemic attack and 11 of these involved aspirin given alone or with dipyridamole [2] (Figure 5.1). They found that use of antiplatelet drugs would reduce the risk of recurrent non-fatal stroke in these patients by about 23%, from about 5% to 4% per year. There was also a significant 14% reduction in vascular death and 22% reduction in occurrence of any vascular event. The beneficial effect of aspirin in preventing ischaemic stroke far outweighs the small increase in risk of intracranial haemorrhage, from 0.2% over two years in controls to 0.3% in patients taking prophylactic aspirin [2].

The recent Second European Stroke Prevention Study (ESPS-2) confirmed the benefit of aspirin (in the very low dose of 25 mg twice daily) in reducing second strokes, with a relative risk reduction compared to placebo of 18%. The risk of stroke or death was reduced by 13%. Aspirin in combination with dipyridamole 200 mg twice daily was apparently even more effective, with a relative risk reduction in stroke of 37% and in stroke or death of 24% [128].

Neither the Dutch TIA Trial, which compared 30 mg and 283 mg [129], nor the UK TIA Trial, which compared 1200 mg and 300 mg [130], reported different rates of haemorrhage, though the wide confidence intervals do not completely rule out the possibility of a dose relationship. Gastrointestinal symptoms were significantly more common in patients taking the higher doses. In the Dutch TIA Trial, minor bleeding (epistaxis, bruising), major gastrointestinal haemorrhage and gastrointestinal symptoms were all more frequently associated with the 283 mg aspirin dose [129]. In the UK TIA Trial of 2435 patients, there were 39 serious gastrointestinal haemorrhages with 1200 mg aspirin, 25 haemorrhages with 300 mg and nine with placebo [130].

As bleeding almost never causes transient ischaemia, aspirin can safely be started as soon as transient ischaemic attack has been diagnosed and without waiting for a CT scan to exclude intracranial haemorrhage. Whilst the Regional Heart Study conducted in the UK in 1992 found that only 39% of patients with a history of stroke were taking daily aspirin [29], a more recent survey conducted by the Stroke Association reported that the number had increased to 86% [131].

In patients with presumed cardioembolic transient ischaemic attack or stroke associated with atrial fibrillation, consideration should be given to anticoagulation rather than aspirin therapy [132].

ACUTE ISCHAEMIC STROKE

Acute cerebral infarction is associated with platelet activation and sludging of the microcirculation in the area around the infarct [133] and increased synthesis of thromboxane [134]. Aspirin given acutely might therefore reasonably be expected to improve the cerebral microcirculation and reduce risk of recurrent stroke, as well as prevent associated myocardial events and venous thrombotic complications of immobility. However, aspirin might also be expected to be associated with increased risk of fatal or disabling intracranial haemorrhage. Consequently, until recently there was wide variation in routine clinical practice [135,136] and a paucity of trial evidence to guide management [137].

This unsatisfactory situation has recently been rectified by the reporting of the small Multicentre Acute Stroke Trial – Italy (MAST-Italy) [138] and of two large trials, the International Stroke Trial (IST) [139] and the Chinese Acute Stroke Trial (CAST) [140]. Both these latter trials tested the benefits and risks of aspirin started within 48 hours of suspected acute ischaemic stroke in a reasonably representative sample totalling over 40 000 hospitalized patients. The IST evaluated a dose of 300 mg of aspirin given daily until discharge or for up to 14 days and CAST compared 160 mg aspirin with matching placebo taken for up to four weeks. The results of the trials were consistent with each other and suggest a relatively small but significant overall benefit of treatment. Thus aspirin started early produces an overall reduction of about nine fewer deaths or non-fatal strokes per 1000 patients treated in the first few weeks and 13 fewer dead or dependent per 1000 after some weeks or months of follow-up. There was a small excess of about two haemorrhagic strokes and of two transfused or fatal extracranial bleeds per 1000 treated, but this was outweighed by the reduction in recurrent ischaemic stroke.

There seems to be little evidence that aspirin affects the brain damage and disability due to the acute event [141] and any benefits of early treatment may simply be explained by the safe extension of secondary prevention into the acute phase [142]. Nevertheless, the overall results would seem to justify starting aspirin as soon as possible in all patients with presumed atherothrombotic stroke, unless there are clear contraindications and provided intracerebral haemorrhage has been excluded by CT scanning. If neuroimaging is not available locally, then aspirin can be justified if haemorrhage is considered unlikely. The initial dose should probably be 300 mg, but a lower maintenance dose is then more appropriate. In patients with presumed cardioembolic stroke secondary to atrial fibrillation, heparin rather than aspirin may be preferred, although heparin was not shown to be more efficacious in this subgroup in IST and certainly increases risk of haemorrhage [143].

VENOUS THROMBOSIS AND PULMONARY EMBOLISM

The Antiplatelet Trialists' Collaboration [144] were able to identify 62 trials of antiplatelet therapy involving 9000 medical and surgical patients in which venous thromboembolism was recorded as an endpoint (Figure 5.3). Their review concluded that antiplatelet therapy reduced the incidence of deep vein thrombosis and pulmonary embolism by about 40% and 60% respectively. However, only one-third of the information included in the analysis was from trials of aspirin, some of the doses used were clearly excessive (up to 3900 mg daily) and the quality of some of the studies was questionable [145]. In general, despite the encouraging results of the systematic overview, aspirin is considered to have very limited value in

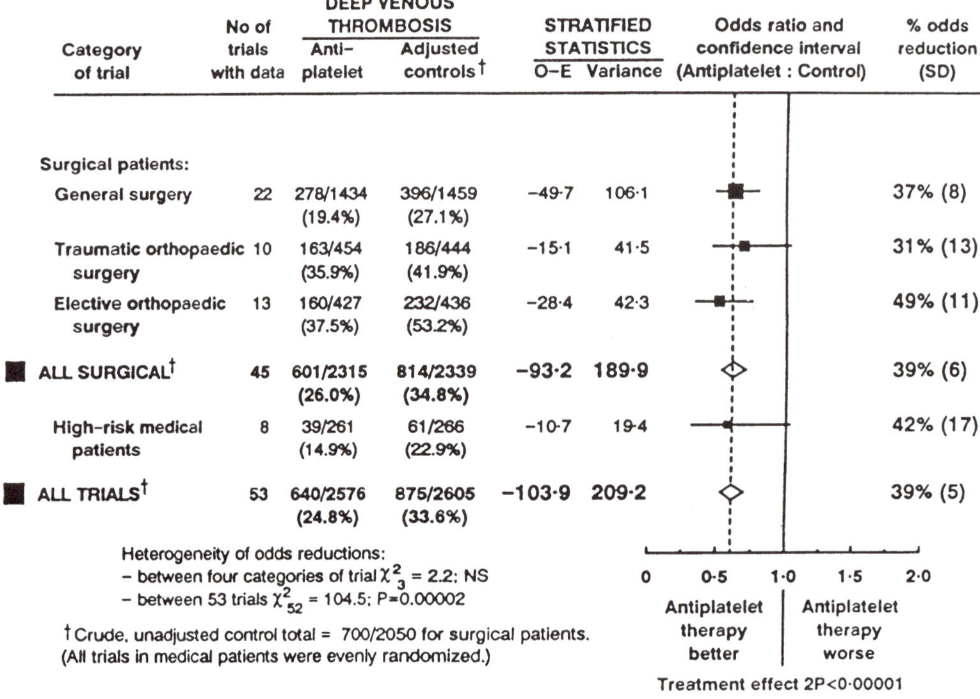

Fig. 5.3 Proportional effects of antiplatelet therapy on numbers of patients in whom deep vein thrombosis was detected by systemic fibrinogen scans or venography or both, after general and orthopaedic (traumatic and elective) surgery and in high-risk medical patients. O − E = observed minus expected (from reference [144] with permission).

preventing deep vein thrombosis [146,147]. It has the advantages of low cost and convenience, but is certainly markedly inferior to use of unfractionated or low molecular weight heparin.

Maintenance of vascular grafts and arterial patency

In the absence of preventive measures, occlusion or reocclusion after coronary or peripheral artery revascularization procedures is common, up to a third of vessels occluding within one year of operation, most within the first few weeks. This may occur subclinically or may produce symptoms of acute arterial obstruction [148]. These patients are at high risk of major clinical vascular events, such as myocardial infarction, stroke or sudden death. It is therefore not surprising that aspirin has proved beneficial after a wide range of vascular surgery, including coronary artery bypass grafts, coronary angioplasty, coronary artery stents, peripheral vessel surgery or angioplasty and formation of a haemodialysis shunt or fistula [149–152].

The Antiplatelet Trialists' Collaboration [153] suggested that antiplatelet therapy (chiefly aspirin alone or aspirin plus dipyridamole) reduced the odds of vascular graft or arterial occlusion after surgery by about 40% while treatment continued, as well as preventing about a quarter of major vascular events (Figure 5.4). In absolute terms, occlusions were prevented in about 40–90 patients per 1000 treated for about seven months after coronary artery procedures and about 19 months after peripheral artery procedures. There was also evidence of substantial benefit in haemodialysis patients with arteriovenous fistulas or shunts, the risk of occlusion being reduced by about two-thirds.

The use of aspirin alone or in combination with other antiplatelet agents or anticoagulant therapy in patients following implantation of coronary or peripheral artery stents is more controversial [154]. Until recently, initial treatment with warfarin for at least a few weeks, together with aspirin started preoperatively and continued indefinitely, was mandatory. Newer techniques of high-pressure balloon inflation with intracoronary ultrasound guidance to ensure correct stent placement and expansion may mean that anticoagulants are not required, with the considerable economic advantages of reducing length of hospital stay [155]. Aspirin alone [156] or aspirin and ticlopidine together [157] have both been claimed to be effective, but evidence from randomized controlled trials is still limited. Numerous studies are in progress to better define optimum management.

Shortly after angioplasty, platelet aggregability is very high and so aspirin therapy must be started at least 24 hours before the procedure to prevent early reocclusion [158]. The timing of initiation of therapy relative to other vascular surgery does not seem to be as crucial, with similar benefit being seen in patients starting treatment before or within 24 hours after operation. However, in a trial of patients undergoing coronary artery bypass surgery, there did appear to be a small excess of bleeding complications when aspirin was started preoperatively [150,159] and initiating aspirin a few hours after operation may reduce risk of perioperative bleeding while still preventing occlusion [160]. Surgery should not be delayed in those patients already taking aspirin, as risk of bleeding is still relatively small.

Whilst the greatest risk of restenosis occurs soon after surgery, when arterial trauma is healing, there is likely to be continued benefit for much longer [161]. Thus, aspirin therapy seems justified in all patients undergoing arterial vascular surgery and should be continued for at least one year and probably indefinitely.

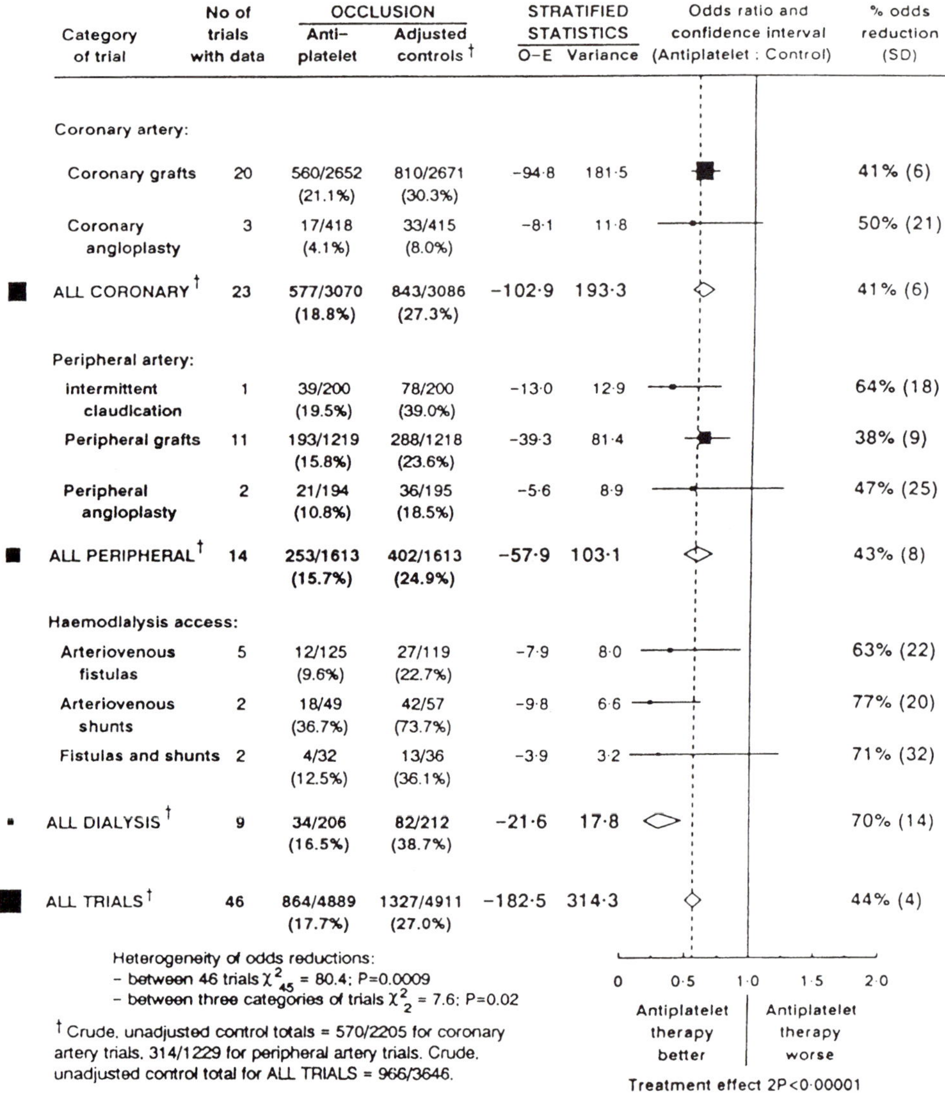

Category of trial	No of trials with data	OCCLUSION: Anti-platelet	OCCLUSION: Adjusted controls†	STRATIFIED STATISTICS: O–E	STRATIFIED STATISTICS: Variance	% odds reduction (SD)
Coronary artery:						
Coronary grafts	20	560/2652 (21.1%)	810/2671 (30.3%)	−94·8	181·5	41% (6)
Coronary angioplasty	3	17/418 (4.1%)	33/415 (8.0%)	−8·1	11·8	50% (21)
ALL CORONARY†	23	577/3070 (18.8%)	843/3086 (27.3%)	−102·9	193·3	41% (6)
Peripheral artery:						
Intermittent claudication	1	39/200 (19.5%)	78/200 (39.0%)	−13·0	12·9	64% (18)
Peripheral grafts	11	193/1219 (15.8%)	288/1218 (23.6%)	−39·3	81·4	38% (9)
Peripheral angioplasty	2	21/194 (10.8%)	36/195 (18.5%)	−5·6	8·9	47% (25)
ALL PERIPHERAL†	14	253/1613 (15.7%)	402/1613 (24.9%)	−57·9	103·1	43% (8)
Haemodialysis access:						
Arteriovenous fistulas	5	12/125 (9.6%)	27/119 (22.7%)	−7·9	8·0	63% (22)
Arteriovenous shunts	2	18/49 (36.7%)	42/57 (73.7%)	−9·8	6·6	77% (20)
Fistulas and shunts	2	4/32 (12.5%)	13/36 (36.1%)	−3·9	3·2	71% (32)
ALL DIALYSIS†	9	34/206 (16.5%)	82/212 (38.7%)	−21·6	17·8	70% (14)
ALL TRIALS†	46	864/4889 (17.7%)	1327/4911 (27.0%)	−182·5	314·3	44% (4)

Fig. 5.4 Proportional effects of antiplatelet therapy on numbers of patients found to have occlusion after coronary artery procedures, peripheral artery procedures or disease and after establishing haemodialysis shunts or fistulas. O − E = observed minus expected (from reference [153] with permission).

What dose?

The optimum dose of aspirin for prevention of thromboembolic disease remains the subject of considerable controversy. This has arisen because of a number of historic factors.

- The clinical development of aspirin for cardiovascular indications was largely driven by the curiosity of the academic community, rather than by the needs of the pharmaceutical industry and licensing authorities. Thus there were none of the standard phase II dose-finding studies prior to embarking on the phase III efficacy trials.

- Aspirin dosage in early trials (900–1500 mg) was based on established analgesic and anti-inflammatory regimens, rather than being based on any preparatory studies of the dose for optimal antiplatelet and antithrombotic effects.
- Early studies of aspirin often involved co-administration of other drugs which required frequent dosing through the day (e.g. three times daily for dipyridamole and four times daily for sulphinpyrazone). For convenience, a standard tablet (e.g. 300 mg) aspirin was used at the same dosing interval.
- The mechanism of action and dose and time dependency of effects of aspirin on COX and endothelial prostacyclin were only fully investigated after clinical efficacy had been established.

The optimum dose of aspirin would fully inhibit platelet COX and so thromboxane production, whilst having a minimal effect on prostacyclin production from the vessel wall. As COX is permanently inhibited by aspirin, the effect of repeated very low doses (less than 100 mg daily) on thromboxane production is cumulative [162]. Thus within 7–10 days, daily doses of only 30–80 mg will result in almost complete suppression of thromboxane synthesis and maximally inhibit platelet aggregation and prolong bleeding time [163]. Higher aspirin doses cannot further suppress thromboxane production. Prostacyclin synthesis is only about 50% inhibited by low-dose aspirin (100 mg or less) and fully recovers within 24 hours [164]. In the setting of acute myocardial infarction or unstable angina and possibly acute ischaemic stroke, a rapid onset of action is desirable and so a single 'loading dose' of aspirin is appropriate. Studies of thromboxane production and platelet function have shown greater than 95% suppression within 30 minutes to an hour of a single oral aspirin dose of 120 mg or more [165,166].

The high doses of aspirin used in early large-scale trials (e.g. greater than 600 mg daily) have now been abandoned, with the results of experimental studies encouraging the adoption of medium doses (e.g. 150–300 mg) for rapid initiation of therapy and low-dose regimens (e.g. 100 mg or less daily) for maintenance therapy [103]. An overview of 10 randomized trials of aspirin in 6171 patients after a transient ischaemic attack or non-disabling stroke found virtually no difference between low (<100 mg daily), medium (300–325 mg daily) or high (>900 mg daily) dose aspirin regimens, with relative risk reductions in vascular events of 13%, 9% and 14% respectively [167]. This equivalence corresponds with the results of direct comparisons made between 300 mg and 1200 mg in the UK TIA Trial [130] and 30 mg and 283 mg in the Dutch TIA Trial [129]. A randomized controlled trial of 50 mg versus 900 mg aspirin daily in patients undergoing angioplasty of aortoiliac and femoropopliteal atherosclerotic lesions showed virtually identical restenosis rates, but significantly fewer serious gastrointestinal side effects in patients taking the lower aspirin dose [168]. To prevent restenosis after coronary angioplasty, a 500 mg dose of aspirin was superior to 100 mg or 40 mg daily [169].

Other studies have suggested specific benefits of particular dose regimens, generally based on experimental work carried out on healthy volunteers, rather than prospective randomized controlled trials. Thus the finding that a 325 mg dose of aspirin inhibited collagen-induced platelet aggregation after exercise to a greater extent than 81 mg led the authors to conclude that the higher dose might be indicated in acute coronary syndromes characterized by plaque rupture with extensive collagen exposure and platelet activation [170]. Another experimental study in healthy volunteers [171] suggested that the optimum antiplatelet dose regimen might be 500 mg of aspirin every two weeks, followed by daily doses of 50 mg. The fortnightly 'loading dose' of aspirin was claimed to diminish the effect of red blood cells in promoting platelet clumping. Others put forward the view that very low doses of 30 mg daily are sufficient, emphasizing the importance of maintaining endogenous myocardial protection by prostacyclin generated by endothelial cells and the advantage of not inducing gastrointestinal ulceration [172].

In practice, the dosage used is still often greatly influenced by the available preparations. Physicians in the USA generally report using higher doses of aspirin for stroke prevention than those in the UK [173] and choice of dosage for acute events reflects the fact that in the USA aspirin is marketed in 325 mg tablets, whereas in Britain the standard dose is 300 mg. The World Health Organization have recommended a dose of 100 mg for long-term secondary prophylaxis and the trend increasingly is to use the lowest proven effective dose. At present, this justifies a daily dose of 50–100 mg.

Conclusions

In patients with established vascular disease, the average annual risk of thromboembolic events in the absence of treatment is about 8–11% [174]. This may be significantly reduced by low-dose aspirin therapy. Thus it is has been estimated that as many as 9000 deaths could be prevented each year in the UK and 100 000 in developed countries by the appropriate use of aspirin and perhaps up to twice this number of non-fatal myocardial infarctions and strokes. The cost has been estimated at about £250 per life saved or about £80 per cardiovascular event avoided [175,176]. Continued evidence of underuse of aspirin is therefore of concern.

Many different pathways are involved in platelet aggregation and, whilst absolute benefits of other currently available antiplatelet drugs over aspirin appear to be small [177–179], there is considerable interest in the possibility that a combination of aspirin with another antiplatelet agent working through a different mechanism might be more effective than aspirin alone [128,180]. The benefit-to-risk ratio of aspirin in combination with oral anticoagulation is also deserving of further study [181].

KEYPOINTS

- In the primary prevention of vascular disease in healthy, low-risk subjects, the potential adverse effects of aspirin will generally outweigh any small benefit.
- Low-dose aspirin is effective in the secondary prevention of vascular events in patients with ischaemic heart disease, stroke and TIA, peripheral vascular disease and after vascular procedures and surgery. It may have a role to play in some patients with diabetes mellitus, thrombocytosis, prosthetic heart valves and atrial fibrillation.
- Meta-analysis suggests potential benefit of aspirin for reducing risk of venous thromboembolism, but it is rarely used clinically.
- The optimum therapeutic dose of aspirin for thromboprophylaxis is debatable, but in most patients is likely to be 50–100 mg daily. When immediate effect is required, such as after acute myocardial infarction or ischaemic stroke, an initial loading dose of at least 150 mg is appropriate.

REFERENCES

1. Antiplatelet Trialists' Collaboration. Secondary prevention of vascular disease by prolonged antiplatelet therapy. Br Med J 1988; 296; 320–331.
2. Antiplatelet Trialists' Collaboration. Collaborative overview of randomised controlled trials of antiplatelet therapy – I: prevention of death, myocardial

infarction, and stroke by prolonged antiplatelet therapy in various categories of patients. Br Med J 1994; 308: 81–106.

3. Collins R, Peto R, Baigent C, Sleight P. Aspirin, heparin and fibrinolytic therapy in suspected acute myocardial infarction. N Engl J Med 1997; 336: 847–860.
4. Boston Collaborative Drug Surveillance Group. Regular aspirin intake and acute myocardial infarction. Br Med J 1974; 1: 440–443.
5. Jick H, Miettinen OS. Regular aspirin use and myocardial infarction. Br Med J 1976; 1: 1057.
6. Hennekens CH, Karlson LK, Rosner B. A case control study of regular aspirin use and coronary deaths. Circulation 1978; 58: 35–38.
7. Hammond EC, Garfinkel L. Aspirin and coronary heart disease: findings of a prospective study. Br Med J 1975; 2: 269–271.
8. Paganini-Hill A, Chao A, Ross RK, Henderson BE. Aspirin use and chronic diseases: a cohort study of the elderly. Br Med J 1989; 299: 1247–1250.
9. Peto R, Gray R, Collins R *et al.* Randomised trial of prophylactic daily aspirin in British male doctors. Br Med J 1988; 296: 313–316.
10. Steering Committee of the Physicians' Health Study Research Group. Final report on the aspirin component of the ongoing Physicians' Health Study. N Engl J Med 1989; 321: 129–135.
11. Manson JE, Stampfer MJ, Colditz GA *et al.* A prospective study of aspirin use and primary prevention of cardiovascular disease in women. JAMA 1991; 266: 565–566.
12. Buring JE, Hennekens CH. The Women's Health Study: summary of the study design. J Myocardial Ischaemia 1992; 4: 27–29.
13. Lamas GA, Pfeffer MA, Hamm P *et al.* Do the results of randomised clinical trials of cardiovascular drugs influence clinical practice? N Engl J Med 1992; 327: 241–247.
14. Ridker PM, Manson JE, Buring JE, Muller JE, Hennekens CH. Circadian variation of acute myocardial infarction and the effect of low-dose aspirin in a randomised trial of physicians. Circulation 1990; 82: 897–902.
15. Tofler GH, Brezinski DA, Schafer AI *et al.* Concurrent morning increase in platelet aggregability and the risk of myocardial infarction and sudden death. N Eng J Med 1987; 316: 1514–1518.
16. Verstraete M. Primary prevention of myocardial infarction and stroke with aspirin. Nutr Metab Cardiovasc Dis 1993; 3: 145–150.
17. CDC Morbidity and Mortality Weekly Report 1997; 46: 498–501.
18. American Academy of Family Physicians, Commission on Public Health and Scientific Affairs. Age Charts for Periodic Health Examination. Kansas City: American Academy of Family Physicians, 1994 (Reprint No. 510).
19. Canadian Task Force on the Periodic Health Examination. Canadian Guide to Clinical Preventive Health Care. Ottawa: Canada Communication Group, 1994: 680–691.
20. Fuster V, Dyken ML, Vokonas PS, Hennekens C. AHA medical/scientific statement: aspirin as a therapeutic agent in cardiovascular disease. Circulation 1993; 87: 659–675.

21. Meade TW. Randomised controlled trial of low-dose warfarin and low-dose aspirin in the primary prevention of ischaemic heart disease. Am J Cardiol 1990; 65: 7C-11C.
22. Meade TW, Roderick PJ, Brennan PJ, Wilkes HC, Kelleher CC. Extracranial bleeding and other symptoms due to low dose aspirin and low intensity oral anticoagulation. Thromb Haemostas 1992; 68: 1–6.
23. Anonymous. New trial aims to identify optimal blood pressure. Scrip 1992; 1729: 26.
24. Thadani U. Management of patients with chronic stable angina at low risk for serious cardiac events. Am J Cardiol 1997; 79: 24–30.
25. Elwood PC, Renaud S, Sharp DS *et al.* Ischaemic heart disease and platelet aggregation: the Caerphilly Collaborative Heart Disease Study. Circulation 1991; 83: 38–44.
26. Juul-Moller S, Edvardsson N, Jahnmatz B, Rosen A, Sorensen S, Omblus R. Double-blind trial of aspirin in primary prevention of myocardial infarction in patients with stable chronic angina pectoris. Lancet 1992; 340: 1421–1425.
27. Ridker PM, Manson JE, Gaziano JM, Buring JE, Hennekens CH. Low-dose aspirin therapy for chronic stable angina: a randomised, placebo-controlled clinical trial. Ann Intern Med 1991: 114: 835–839.
28. Manson JE, Grobbee DE, Stampfer MJ *et al.* Aspirin in the primary prevention of angina pectoris in a randomised controlled trial of US physicians. Am J Med 1990; 89: 772–776.
29. McCallum AK, Whincup PH, Morris RW *et al.* Aspirin use in middle aged men with cardiovascular disease: are opportunities being missed? Br J Gen Pract 1997; 47: 417–421.
30. King R, Denne J. Audit suggests that use of aspirin in coronary heart disease is rising. Br Med J 1995; 311: 1504.
31. Carney AJ, Carney TA. Use of aspirin in secondary prevention of coronary heart disease is rising. Br Med J 1996; 312: 846.
32. Batty GM, Oborne CA, Close JCT, Swift CG, Jackson SHD. Appropriate use of aspirin in patients aged greater than 65 years with ischaemic heart disease. Age Ageing 1997; 26 (suppl 3): 7.
33. McCartney P, Macdowall W, Thorogood M. A randomised controlled trial of feedback to general practitioners of their prophylactic aspirin prescribing. Br Med J 1997; 315: 35–36.
34. Goldhaber SZ, Manson JA, Stampfer MJ *et al.* Low dose aspirin and subsequent peripheral arterial surgery in the Physicians' Health Study. Lancet 1992; 340: 143–145.
35. Walters TK, Mitchell DC, Wood RF. Low-dose aspirin fails to inhibit increased platelet reactivity in patients with peripheral vascular disease. Br J Surg 1993; 80: 1266–1268.
36. Clagett GP, Graor RA, Salzman EW. Antithrombotic therapy in peripheral arterial occlusive disease. Chest 1992; 102 (suppl): 516S-528S.

37. Wolf PA, Dawber TR, Thomas HE, Kannel WB. Epidemiologic assessment of chronic atrial fibrillation and risk of stroke: the Framingham Study. Neurology 1978; 28: 973–977.
38. The Stroke Prevention in Atrial Fibrillation Investigators. Predictors of thromboembolism in atrial fibrillation: 1. Clinical features of patients at risk. Ann Intern Med 1992; 116: 1–5.
39. The Stroke Prevention in Atrial Fibrillation Investigators. Predictors of thromboembolism in atrial fibrillation: II. Echocardiographic features of patients at risk. Ann Intern Med 1992; 116: 6–12.
40. Laupacis A, Albers GW, Dalen JE *et al.* Antithrombotic therapy in atrial fibrillation. Chest 1995; 108: 352S–360S.
41. Petersen P, Boysen G, Godtfredsen J, Andersen ED, Andersen B. Placebo-controlled, randomised trial of warfarin and aspirin for the prevention of thromboembolic complication in chronic atrial fibrillation. The Copenhagen AFASAK Study. Lancet 1989; 1: 175–179.
42. Stroke Prevention in Atrial Fibrillation Investigators. Stroke Prevention in Atrial Fibrillation Study: final results. Circulation 1991; 84: 527–539.
43. Stroke Prevention in Atrial Fibrillation Investigators. Warfarin versus aspirin for prevention of thromboembolism in atrial fibrillation: Stroke Prevention in Atrial Fibrillation II Study. Lancet 1994; 343: 687–691.
44. Stroke Prevention in Atrial Fibrillation Investigators. Adjusted-dose warfarin versus low-intensity fixed-dose warfarin plus aspirin for high-risk patients with atrial fibrillation: Stroke Prevention in Atrial Fibrillation III randomised clinical trial. Lancet 1996; 348: 633–638.
45. The Boston Area Anticoagulation Trial for Atrial Fibrillation Investigators. The effect of low-dose warfarin on the risk of stroke in patients with nonrheumatic atrial fibrillation. N Engl J Med 1990; 323: 1505–1511.
46. Atrial Fibrillation Investigators. Risk factors for stroke and efficacy of antithrombotic therapy in atrial fibrillation: analysis of pooled data from five randomised controlled trials. Arch Intern Med 1994; 154: 1449–1457.
47. Lip GYH, Lowe GDO. Warfarin and aspirin as thromboprophylaxis in atrial fibrillation. Br J Clin Pharmacol 1996; 41: 368–379.
48. Atrial Fibrillation Investigators. The efficacy of aspirin in patients with atrial fibrillation. Analysis of pooled data from 3 randomised trials. Arch Intern Med 1997; 157: 1237–1240.
49. EAFT (European Atrial Fibrillation Trial) Study Group. Secondary prevention in non-rheumatic atrial fibrillation after transient ischaemic attack or minor stroke. Lancet 1993; 342: 1255–1262.
50. Lip GYH, Lip PL, Zaritis J *et al.* Fibrin D-dimer and beta-thromboglobulin as markers of thrombogenesis and platelet activation in atrial fibrillation: effects of introducing ultra-low dose warfarin and aspirin. Circulation 1996; 94: 425–431.
51. More RS, Chauhan A. Anti-thrombotic therapy for non-rheumatic atrial fibrillation. Q J Med 1996; 89: 409–414.
52. Lancaster T, Mant J, Singer DE. Stroke prevention in atrial fibrillation. Br Med J 1997; 314: 1563–1564.

53. Fuller JH, Shipley MJ, Rose G, Jarret RJ, Keen H. Mortality from coronary heart disease and stroke in relation to the degree of glycaemia: the Whitehall Study. Br Med J 1983; 287: 867–870.
54. Davi G, Catalano I, Averna M *et al.* Thromboxane biosynthesis and platelet function in type II diabetes mellitus. N Engl J Med 1990; 322: 1769–1774.
55. Mori TA, Vandongen R, Douglas AJ, McCulloch RK, Burke V. Differential effect of aspirin on platelet aggregation in IDDM. Diabetes 1992; 41: 261–266.
56. De La Cruz JP, Moreno A, Munoz M, Garcia Campos JM, Sanchez De La Cuesta F. Effect of aspirin plus dipyridamole on the retinal vascular pattern in experimental diabetes mellitus. J Pharmacol Exp Ther 1997; 280: 454–459.
57. Moel DI, Safirstein RL, McEvoy RC, Hsueh W. Effect of aspirin on experimental diabetic nephropathy. J Lab Clin Med 1987; 110: 300–307.
58. Klein BE, Klein R, Moss SE. Is aspirin usage associated with diabetic retinopathy? Diabetes Care 1987; 10: 600–603.
59. Collwell JA, Bingham SF, Abraira C *et al.* The Cooperative Study Group: Veterans Administration Cooperative Study on antiplatelet agents in diabetic patients after amputation for gangrene: II effects of aspirin and dipyridamole on atherosclerotic vascular disease rates. Diabetes Care 1986; 9: 140–148.
60. Collwell JA, Bingham SF, Abraira C *et al.* The Cooperative Study Group: Veterans Administration Cooperative Study on antiplatelet agents in diabetic patients after amputation for gangrene: unobserved, sudden, and unexpected deaths. J Diabetic Compl 1989; 3: 191–197.
61. DAMAD Study Group. Effect of aspirin alone and aspirin plus dipyridamole in early diabetic retinopathy. A multicenter randomized controlled clinical trial. Diabetes 1989; 38: 491–498.
62. ETDRS Investigators. Aspirin effects on mortality and morbidity in patients with diabetes mellitus: the Early Treatment Diabetic Retinopathy Study report 14. JAMA 1992; 268: 1292–1300.
63. Chew EY, Klein ML, Murphy RP, Remaley NA, Ferris FL 3rd. Effects of aspirin on vitreous/preretinal haemorrhage in patients with diabetes mellitus. Early Treatment Diabetic Retinopathy Study report no. 20. Arch Ophthalmol 1995; 113: 52–55.
64. Barnett AH, Letherdale BA, Polak A *et al.* Specific thromboxane synthetase inhibition and albumin excretion rate in insulin-dependent diabetes. Lancet 1984; i: 1322–1324.
65. Donadio JV Jr, Ilstrup DM, Holley KE, Romero JC. Platelet-inhibitor treatment of diabetic nephropathy: a 10 year prospective study. Mayo Clin Proc 1988; 63: 3–15.
66. Contreras I, Reiser KM, Martinez M *et al.* Effects of aspirin or basic amino acids on collagen cross-links and complications in NIDDM. Diabetes Care 1997; 20: 832–835.
67. Yudkin JS. Assessing the evidence on aspirin in diabetes mellitus. Gemini or libra, lumping or splitting, surrogate or hard, interventionist or nihilist. Diabetologia 1996; 39: 1407–1408.

68. Turpie AGG, Gent M, Laupacis A *et al.* A comparison of aspirin with placebo in patients treated with warfarin after heart-valve replacement. N Engl J Med 1993; 329: 524–529.
69. Landolfi R, Patrono C. Aspirin in polycythaemia vera and essential thrombocythaemia: current facts and perspectives. Leuk Lymphoma 1996; 22 (suppl 1): 83–86.
70. Landolfi R, Ciabattoni G, Patrignani P *et al.* Increased thromboxane biosynthesis in patients with polycythaemia vera: evidence for aspirin-suppressible platelet activation in vivo. Blood 1993; 82: 1377–1378.
71. Landolfi R, Marchioli R, Patrono C. Mechanisms of bleeding and thrombosis in myeloproliferative disorders. Thromb Haemostas 1997; 78: 617–621
72. Tartaglia A, Goldber J, Berk P, Wasserman L. Adverse effects of antiaggregating platelet therapy in the treatment of polycythaemia vera. Semin Hematol 1986; 23: 172–176.
73. Gilbert HS, Hanna MM. Low dose aspirin/low haematocrit regimen: a safe and efficacious treatment for polycythaemia vera. Blood 1986; 68: 770a.
74. Anonymous. Low dose aspirin in polycythaemia vera: a pilot study. Gruppo Italiano Studio Policitemia – GISP. Br J Haematol 1997; 97: 453–456.
75. Van Genderen PJJ, Michiels J, Van Strik R, Lindemans J, Van Vliet HHDM. Platelet consumption in thrombocythaemia complicated by erythromelalgia: reversal by aspirin. Thromb Haemostas 1995; 73: 210–214.
76. Michiels J, Koudstaal P, Mulder A, Van Vliet H. Transient neurologic and ocular manifestation in primary thrombocythaemia. Neurology 1993; 43: 1107–1110.
77. Jorgensen HP, Sondergard J. Pathogenesis of erythromelalgia. Arch Dermatol 1978; 114: 112–114.
78. Van Genderen P, Michiels J. Erythromelalgic, thrombotic and haemorrhagic manifestations of thrombocythaemia. Press Med 1994; 23: 73–77.
79. Van Genderen PJ, Mulder PG, Waleboer M, Van De Moesdijik D, Michiels JJ. Prevention and treatment of thrombotic complications in essential thrombocythaemia: efficacy and safety of aspirin. Br J Haematol 1997; 97: 179–184.
80. Theroux P, Lator JG, Leger-Gauthier C, DeLara J. Fibrinopeptide A and platelet factor levels in unstable angina. Circulation 1987; 75: 156–162.
81. Chesebro JH, Fuster V. Thrombosis in unstable angina. N Engl J Med 1992; 327: 192–194.
82. Theroux P, Ouimet H, McCans J *et al.* Aspirin, heparin, or both to treat acute unstable angina. N Engl J Med 1988; 319: 1105–1111.
83. Lewis HD Jr, Davis JW, Archibald DG *et al.* Prospective effects of aspirin against acute myocardial infarction and death in men with unstable angina: results of a Veterans Administration Cooperative Study. N Engl J Med 1983; 309: 396–403.
84. Cairns JA, Gent M, Singer J *et al.* Aspirin, sulfinpyrazone, or both in unstable angina: results of a Canadian multicenter trial. N Engl J Med 1985; 313: 1369–1375.

85. The RISC Group. Risk of myocardial infarction and death during treatment with low dose aspirin and intravenous heparin in men with unstable coronary artery disease. Lancet 1990; 336: 827–830.
86. Wallentin LC, Research Group on Instability in Coronary Artery Disease in Southeast Sweden. Aspirin (75mg/day) after an episode of unstable coronary artery disease: long term effects on the risk for myocardial infarction, occurrence of severe angina and the need for revascularisation. J Am Coll Cardiol 1991; 18: 1587–1593.
87. Nyman I, Larsson H, Wallentin L, Research Group on Instability in Coronary Artery Disease in Southeast Sweden. Prevention of serious cardiac events by low-dose aspirin in patients with silent myocardial ischaemia. Lancet 1992; 340: 497–501.
88. Theroux P, Waters D, Lam J, Juneau M, McCans J. Reactivation of unstable angina after the discontinuation of heparin. N Engl J Med 1992: 327: 141–145.
89. ISIS-2 (Second International Study of Infarct Survival) Collaborative Group. Randomised trial of intravenous streptokinase, oral aspirin, both, or neither among 17,187 cases of suspected acute myocardial infarction. ISIS-2. Lancet 1988; 2: 349–360.
90. Baigent C, Collins R. ISIS-2: 4–year mortality follow up of 17,187 patients after fibrinolytic and antiplatelet therapy in suspected acute myocardial infarction. Circulation 1993: 88 (suppl 1): I-291.
91. Roux S, Christeller E, Ludin E. Effects of aspirin on coronary reocclusion and recurrent ischaemia after thrombolysis: a meta-analysis. J Am Coll Cardiol 1992; 19: 671–677.
92. Brown N, Young T, Gray D, Skene AM, Hampton JR. Inpatient deaths from myocardial infarction, 1982–92: analysis of data in the Nottingham Heart Attack Register. Br Med J 1997; 315: 159–164.
93. Hennekens CH, Jonas MA, Buring JE. The benefits of aspirin in acute myocardial infarction: still a well-kept secret in the US. Arch Intern Med 1994; 1: 37–39.
94. McLaughlin TJ, Soumerai SB, Willison DJ *et al.* Adherence to national guidelines for drug treatment of suspected acute myocardial infarction: evidence for undertreatment in women and the elderly. Arch Intern Med 1996; 156: 799–805.
95. Norris RM, White HD, Cross DB *et al.* Aspirin does not improve early arterial patency after streptokinase treatment for acute myocardial infarction. Br Heart J 1993; 69: 492–495.
96. Dorn GW, Liel N, Trask JL *et al.* Increased platelet thromboxane A2/ prostaglandin H2 receptors in patients with acute myocardial infarction. Circulation 1990; 81: 212–218.
97. Jaschonek K, Karsch KR, Weisenberger H *et al.* Platelet prostacyclin binding in coronary artery disease. J Am Coll Med 1986; 8: 259–266.
98. Fitzgerald DJ, Catella F, Roy L, Fitzgerald GA. Marked platelet activation in vivo after intravenous streptokinase in patients with acute myocardial infarction. Circulation 1988; 77: 142–150.

99. Rudd MA, George D, Amarante P, Vaughan DE, Loscalzo J. Temporal effects of thrombolytic agents on platelet function in vivo and their modulation by prostaglandins. Circ Res 1990; 67: 1175–1181.
100. Rasmanis G, Vestequist O, Green K, Edhag O, Henriksson P. Evidence of increased platelet activation after thrombolysis in patients with acute myocardial infarction. Br Heart J 1992; 68: 374–376.
101. Meijer A, Verheugt FW, Werter CJ *et al.* Aspirin versus coumadin in the prevention of reocclusion and recurrent ischaemia after successful thrombolysis: a placebo controlled angiographic study. Results of the APRICOT study. Circulation 1993; 87: 1524–1530.
102. Julian DG, Chamberlain DA, Pocock SJ for the AFTER Study Group. A comparison of aspirin and anticoagulation following thrombolysis for myocardial infarction (the AFTER Study): a multicentre unblinded randomised clinical trial. Br Med J 1996; 313: 1429–1431.
103. Kearon C, Hirsh J. Optimal dose for starting and maintaining low dose aspirin. Arch Intern Med 1993; 153: 700–702.
104. Elwood P. Gastric safety and enteric-coated aspirin. Lancet 1997; 349: 432.
105. Brecker SJD. Early aspirin in myocardial infarction. Lancet 1990; 335: 923.
106. NHS Executive. The Health of the Nation. Assessing the Options: CHD/Stroke. Target Effectiveness and Cost Effectiveness of Interventions to Reduce Coronary Heart Disease and Stroke Mortality. London: Department of Health, 1995.
107. Moher M, Johnson N. Use of aspirin by general practitioners in suspected acute myocardial infarction. Br Med J 1995; 308: 760.
108. Wylie HR, Dunn FG. Pre-hospital opiate and aspirin administration in patients with suspected myocardial infarction. Br Med J 1994; 308: 760.
109. Moher M, Moher D, Havelock P. Survey of whether general practitioners carry aspirin in their doctor's bag. Br Med J 1994; 308: 761–762.
110. Saketkhou BB, Conte FJ, Norris M *et al.* Emergency department use of aspirin in patients with possible acute myocardial infarction. Ann Intern Med 1997; 127: 126–129.
111. Elwood PC, Cochrane AJ, Burr ML *et al.* A randomised controlled trial of acetylsalicylic acid in the secondary prevention of mortality from myocardial infarction. Br Med J 1974; 1: 436–440.
112. Elwood PC, Sweetnam PM. Aspirin and secondary mortality after myocardial infarction. Lancet 1979; ii: 1313–1315.
113. Elwood PC, Williams WO. A randomised controlled trial of aspirin in the prevention of early mortality in myocardial infarction. J Roy Coll Gen Pract 1979; 29: 413–416.
114. Coronary Drug Project Research Group (CDP-A). Aspirin in coronary heart disease. J Chron Dis 1976; 29: 625–642.
115. Breddin K, Loew D, Lechner K, Uberla K, Walter E. Secondary prevention of myocardial infarction: a comparison of acetylsalicylic acid, placebo and phenprocoumon. Haemostasis 1980; 9: 325–344.

116. Aspirin Myocardial Infarction Study Research Group (AMIS Group). A randomised controlled trial of aspirin in persons recovered from myocardial infarction. JAMA 1980; 243: 661–669.
117. Persantine–Aspirin Reinfarction Study Research Group (PARIS-1). Persantine and aspirin in coronary heart disease. Circulation 1980; 62: 449–461.
118. Elwood PC. Aspirin in the prevention of myocardial infarction. Current status. Drugs 1984: 28: 1–5.
119. Klimt CR, Knatterud GL, Stamler J, Meier P (PARIS-II Investigators Group). Persantine–aspirin reinfarction study. Part II. Secondary coronary prevention with persantine and aspirin. J Am Coll Cardiol 1986; 7: 251–269.
120. ASPIRE Steering Group. A British Cardiac Society survey of the potential for the secondary prevention of coronary disease: ASPIRE (Action on Secondary Prevention through Intervention to Reduce Events). Heart 1996: 75: 334–342.
121. Schwartz LM, Fisher ES, Tosteson NA *et al.* Treatment and health outcomes of women and men in a cohort with coronary artery disease. Arch Intern Med 1997; 157: 1545–1551.
122. Burn J, Dennis MS, Bamford J *et al.* Long term risk of recurrent stroke after a first ever stroke. The Oxfordshire Community Stroke Project. Stroke 1994; 25: 333–337.
123. Dennis M, Bamford J, Sandercock P, Warlow C. Prognosis of transient ischaemic attacks in the Oxfordshire Community Stroke Project. Stroke 1990; 21: 848–853.
124. Fields WS, Lemak NA, Frankowski RF, Hardy RJ. Controlled trial of aspirin in cerebral ischaemia. Stroke 1977; 8: 301–316.
125. Fields WS, Lemak NA, Frankowski RF, Hardy RJ. Controlled trial of aspirin in cerebral ischaemia. Part II. Surgical group. Stroke 1978; 9: 309–318.
126. ESPS Group. The European Stroke Prevention Study (ESPS). Principal end-points. Lancet 1987; ii: 1351–1354.
127. SALT Collaborative Group. Swedish Aspirin Low-dose Trial (SALT) of 75mg aspirin as secondary prophylaxis after cerebrovascular events. Lancet 1991; 338: 1345–1349.
128. Diener H, Cunha L, Forbes C *et al.* European Stroke Prevention Study 2. Dipyridamole and acetylsalicylic acid in the secondary prevention of stroke. J Neurol Sci 1996; 143: 1–13.
129. Dutch TIA Trial Study Group. A comparison of two doses of aspirin (30mg vs. 283mg a day) in patients after a transient ischaemic attack or minor ischaemic stroke. N Engl J Med 1991; 325: 1261–1266.
130. Farrell B, Godwin J, Richards S, Warlow C. The United Kingdom transient ischaemic attack (UK-TIA) aspirin trial: final results. J Neurol Neurosurg Psychiatr 1991; 54: 1044–1054.
131. Anonymous. Are Opportunities Being Missed? Preventing Recurrent Strokes. London: Stroke Association, 1997: 1–20.
132. Barnett HJM, Eliasziw M, Meldrum HE. Drugs and surgery in the prevention of ischaemic stroke. N Engl J Med 1995; 332: 238–248.

133. Van Kooten F, Ciabattoni G, Patrono C *et al.* Evidence for episodic platelet activation in stroke. Stroke 1994; 25: 278–281.
134. Koudstaal P, Ciabattoni G, Van Gijn J *et al.* Increased thromboxane biosynthesis in patients with acute cerebral ischaemia. Stroke 1993; 24: 219–223.
135. Kent J, Bamford J. Consultant views on the use of aspirin in acute cerebrovascular disease: implications for clinical trials. Postgrad Med J 1994; 70: 185–187.
136. Marsh EE, Adams HP, Biller J *et al.* Use of antithrombotic drugs in the treatment of acute ischaemic stroke. A survey of neurologists in practice in the United States. Neurology 1989; 39: 1631–1634.
137. Sandercock PAG, Van Der Belt A, Lindley R, Slattery J. Antithrombotic therapy in acute ischaemic stroke: an overview of the completed randomised trials. J Neurol Neurosurg Psychiat 1993; 56: 17–25.
138. Multicentre Acute Stroke Trial – Italy (MAST-I) Group. Randomised controlled trial of streptokinase, aspirin, and combination of both in acute ischaemic stroke. Lancet 1995; 346: 1509–1514.
139. International Stroke Trial (IST) Collaborative Group. The International Stroke Trial (IST): a randomised trial of aspirin, subcutaneous heparin, both, or neither among 19435 patients with acute ischaemic stroke. Lancet 1997; 349: 1569–1581.
140. CAST (Chinese Acute Stroke Trial) Collaborative Group. CAST: randomised placebo-controlled trial of early aspirin use in 20000 patients with acute ischaemic stroke. Lancet 1997; 349: 1641–1649.
141. De Keyser J, Herroelen L, De Klippel N. Early outcome in acute ischaemic stroke is not influenced by the prophylactic use of low-dose aspirin. J Neurol Sci 1997; 145: 93–96.
142. Barer D. Interpretation of IST and CAST stroke trials. Lancet 1997: 350: 443–444.
143. Bousser MG. Aspirin or heparin immediately after a stroke? Lancet 1997; 349: 1564–1565.
144. Antiplatelet Trialists' Collaboration. Collaborative overview of randomised controlled trials of antiplatelet therapy - III: reduction in venous thrombosis and pulmonary embolism by antiplatelet prophylaxis among surgical and medical patients. Br Med J 1994; 308: 235–246.
145. Hirsh J, Dalen JE, Fuster V, Harker LB, Salzman EW. Aspirin and other antiplatelet drugs: the relationship between dose, effectiveness, and side effects. Chest 1992; 102 (suppl): 327S–336S.
146. Thromboembolic Risk Factors (THRIFT) Consensus Group. Risk of and prophylaxis for venous thromboembolism in hospital patients. Br Med J 1992; 305: 567–574.
147. Pineo GF, Hull RD. Prevention and treatment of venous thromboembolism. Drugs 1996; 52: 71–92.
148. Fuster V, Chesebro JH. Role of platelets and platelet inhibitors in aortocoronary vein graft disease. Circulation 1986; 73: 227–232.

149. Lorenz RL, Schacky CV, Weber M *et al.* Improved aortocoronary artery bypass patency by low-dose aspirin (100mg daily). Lancet 1984: i: 1261–1264.
150. Goldman S, Copeland J, Moritz T *et al.* Saphenous vein graft patency 1 year after coronary artery bypass surgery and effects of antiplatelet therapy: results of a Veterans Administration Cooperative Study. Circulation 1989; 80: 1190–1197.
151. Gavaghan TP, Gebski V, Baron DW. Immediate postoperative aspirin improves vein graft patency early and late after coronary artery bypass graft surgery. A placebo-controlled, randomised study. Circulation 1991; 83: 1526–1533.
152. Stein PD, Dalen JE, Goldman S *et al.* Antithrombotic therapy in patients with saphenous vein and internal mammary artery bypass grafts following transluminal coronary angioplasty. Chest 1995; 108: 424S–430S.
153. Antiplatelet Trialists' Collaboration. Collaborative overview of randomised controlled trials of antiplatelet therapy - II: maintenance of vascular graft or arterial patency by antiplatelet therapy. Br Med J 1994; 308: 159–168.
154. Spinler S, Cheng J. Antithrombotic therapy after intracoronary stenting. Pharmacotherapy 1997; 17: 74–90.
155. Stephens NG, Ludman PF, Petch MC, Schofield PM, Shapiro M. Changing from intensive anticoagulation to treatment with aspirin alone for coronary stents: the experience of one centre in the United Kingdom. Heart 1996; 76: 238–242.
156. Albiero R, Hall P, Itoh A *et al.* Results of a consecutive series of patients receiving only antiplatelet therapy after optimized stent implantation. Comparison of aspirin alone versus combined ticlopidine and aspirin therapy. Circulation 1997; 95: 1145–1156.
157. Hobson AG, Sowinski KM. Ticlopidine and aspirin therapy following implantation of coronary artery stents. Ann Pharmacother 1997; 31: 770–772.
158. Schwartz L, Bourassa MG, Lesperance J *et al.* Aspirin and dipyridamole in the prevention of restenosis after percutaneous transluminal coronary angioplasty. N Engl J Med 1988; 318: 1714–1719.
159. Goldman S, Copeland JG, Moritz T *et al.* Starting aspirin therapy after operation: effects on early graft patency. Circulation 1991; 84: 520–526.
160. Israel DH, Adams PC, Stein B, Chesebro JH, Fuster V. Antithrombotic therapy in the coronary vein graft patient. Clin Cardiol 1991; 14: 283–295.
161. Taylor RR, Gibbons FA, Cope GD *et al.* Effects of low-dose aspirin on restenosis after coronary angioplasty. Am J Cardiol 1991; 68: 874–878.
162. Patrignani P, Filabozzi P, Patrono C. Selective cumulative inhibition of platelet thromboxane production by low dose aspirin in healthy subjects. J Clin Invest 1982; 69: 1366–1372.
163. De Caterina R, Giannessi D, Boem A *et al.* Equal antiplatelet effects of aspirin 50 or 324 mg/day in patients after acute myocardial infarction. Thromb Haemostas 1985; 54: 528–532.

164. Weksler BB, Tack-Goldman K, Subramanian VA, Gay WA Jr. Cumulative inhibitory effect of low-dose aspirin on vascular prostacyclin and platelet thromboxane production in patients with atherosclerosis. Circulation 1985; 71: 332–340.
165. Dabaghi SF, Kamat SG, Payne J *et al.* Effects of low-dose aspirin on in-vitro platelet aggregation in the early minutes after ingestion in normal subjects. Am J Cardiol 1994; 74: 720–723.
166. Reilly IAG, Fitzgerald GA. Inhibition of thromboxane formation in vivo and ex vivo: implications for therapy with platelet inhibitory drugs. Blood 1987; 68: 180–186.
167. Algra A, Van Gijn J. Aspirin at any dose above 30mg offers only modest protection after cerebral ischaemia. J Neurol Neurosurg Psychiatr 1996; 60: 197–199.
168. Ranke C, Creutzig A, Luska G *et al.* Controlled trial of high versus low dose aspirin treatment after percutaneous transluminal angioplasty in patients with peripheral vascular disease. Clin Invest 1994; 72: 673–680.
169. Darius H, Sellig S, Belz GG, Darius BN. Aspirin 500 mg/d is superior to 100 and 40 mg/d for prevention of restenosis following PTCA. Circulation 1994; 90: I-651.
170. Feng D, McKenna C, Murillo J *et al.* Effect of aspirin dosage and enteric coating on platelet reactivity. Am J Cardiol 1997; 80: 189–193.
171. Santos MT, Valles J, Aznar J *et al.* Prothrombotic effect of erythrocytes on platelet reactivity. Reduction by aspirin. Circulation 1997; 95: 63–68.
172. Forster W, Parratt JR. The case of low-dose aspirin for the prevention of myocardial infarction: but how low is low? Cardiovasc Drugs Ther 1997; 10: 727–734.
173. Goldstein LB, Farmer A, Matchar DB. Primary-care physician reported secondary and tertiary stroke prevention practices. A comparison between the United States and the United Kingdom. Stroke 1997; 28: 746–751.
174. Patrono C. Aspirin as an antiplatelet drug. N Engl J Med 1994; 330: 1287–1294.
175. Elwood P, Hughes C. Aspirin and Cardiovascular Disease. Cardiff: Crystal Print, 1997: 1–20.
176. Medical Research Council. Press release: life-saving aspirin. January 7 1994.
177. Leon MB, Baim DS, Gordon P *et al.* Clinical and angiographic results from the Stent Anticoagulation Regimen Study (STARS). Circulation 1996; 94: I–685.
178. CAPRIE Steering Committee. A randomised, blinded trial of clopidogrel versus aspirin in patients at risk of ischaemic events (CAPRIE). Lancet 1996; 348: 1329–1339.
179. Alexander JH, Harrington RA. Antiplatelet and antithrombin therapies in the acute coronary syndromes. Curr Opin Cardiol 1997; 12: 427–437.
180. Born GVR, Collins R. Aspirin versus clopidogrel: the wrong question? Lancet 1997; 349: 806–807.
181. Hafner J, De Moerloose P, Bounameaux H. Oral anticoagulation alone or in combination with aspirin: risks and benefits. Vasa 1996; 25: 1–12.

Miscellaneous therapeutic uses

Introduction

Findings from numerous epidemiological studies and better understanding of the molecular, biochemical and pharmacological mechanisms underlying the action of aspirin have led to insights into a number of potential new therapeutic uses. The ready availability, widespread familiarity, low cost and relative safety of aspirin make it an attractive drug to many clinicians and public health physicians and to patients themselves. However, the pharmaceutical industry understandably no longer gives any research priority to exploring its clinical potential and new developments tend to occur haphazardly. Consequently the volume and strength of the evidence in each indication may not necessarily be representative of its true clinical importance.

Whilst the support for a routine protective role of aspirin in pregnancy has recently faded, evidence of its preventive potential in colorectal cancer, cataract and cognitive decline is increasing. Other possible clinical indications continue to emerge.

Prevention of colorectal cancer

There is an increasing body of epidemiological and experimental evidence that aspirin may inhibit the occurrence or growth of gastrointestinal tumours, especially of the large bowel. However, no randomized controlled trial has yet confirmed the protective effect of aspirin.

EXPERIMENTAL EVIDENCE

Chemically induced cancer in the rat colon is very similar to human colon cancer except for less tendency to metastasize. In rodent models, high doses of aspirin have been shown to reduce the incidence, numbers and size of adenomas and carcinomas of the colon induced by a variety of chemical carcinogens [1–5]. Similar results have also been obtained with numerous other COX inhibitors [5] and these studies suggest that the effect may be at least in part due to suppression of tumour promotion associated with inhibition of COX-induced prostaglandin synthesis. Tumours resume growth after withdrawal of the COX inhibitor. However, it has also been shown that treatment of rats with aspirin or indomethacin using doses which suppress prostaglandin production also increases uptake of tritiated thymidine into intestinal mucosal DNA [6].

Several possible mechanisms have been suggested by which aspirin might exert its apparent protective effect in colon cancer. Inhibition of COX activity and prostaglandin synthesis is generally thought to be most important [7].

Human colorectal cancers and some adenomas have been shown to produce higher levels of prostaglandins, especially of the E series, than normal colon and it is suggested that these might accelerate tumour growth and the tendency to become invasive [8–11]. Suppression of prostaglandin production by aspirin might be expected to have a significant influence on tumour cell growth and proliferation.

Mitogens and growth factors such as transforming growth factor alpha (TGF-alpha), as well as tumour-associated increase in prostaglandin production, appear to be specifically associated with increased expression of the COX-2 gene. Significantly higher levels of COX-2 messenger RNA and protein are found in intestinal tumours in humans and rodents than in adjacent normal colorectal tissue, with upregulation from two- to 50-fold in 85–90% of colorectal adenocarcinomas [12,13]. In contrast, COX-1 messenger RNA levels are not elevated in carcinoma. Treatment with a selective COX-2 inhibitor of mice implanted with transformed human colon cancer (HCA-7) cells which constitutively express high levels of COX-2 protein reduced tumour formation by 85–90% [14]. Thus the protective effect of aspirin may be specifically related to its ability to inhibit COX-2.

Inhibition of COX-1 and COX-2 by aspirin leads to increased production of the leukotriene, 15-hydroxyeicosatetraenoic acid (15-HETE). This possesses antimitogenic properties through its inhibitory effect on leukotriene B4, which may activate carcinogens by oxygenation of phospholipids to hydroperoxides [15]. Aspirin may also reduce oxidative stress induced by the COX pathway. For example, the activation of carcinogens such as malondialdehyde is dependent on COX and metabolism by COX of bile acids in the gut wall may stimulate production of oxygen free radicals in the intestinal mucosa [10].

Other effects of aspirin not related to inhibition of prostaglandin synthesis have also been suggested as relevant. These include its ability to suppress the activation of NF-kappa B [16] and activator protein 1 (AP-1)

[17], which regulate the transcription of key anti-inflammatory and immune response genes, interference with cytokine expression in colonic epithelial cells [18] and inhibition of progression of the normal cell cycle and induction of apoptosis in intestinal mucosal cells [19]. A homoeopathic explanation has even been proposed relating to the ability of aspirin to induce the symptoms suggestive of colon cancer [20].

EPIDEMIOLOGICAL STUDIES

The first study which showed an inverse relationship between regular use of aspirin and colorectal cancer was a population-based case control study conducted in Melbourne, Australia [21]. This reported about a 40–70% lower incidence of colorectal cancer among aspirin users. Subsequent case control studies carried out mainly among hospital-based populations in the US generally confirmed the relationship, with about a 50% lower risk of incident colorectal cancer among people regularly using aspirin compared to non-users [22–25]. A lower than average risk of developing sporadic colorectal adenomas, the usual precursors of non-familial cancers, has also been shown [25,26].

In some studies there was evidence of risk decreasing with duration of exposure and cumulative dose of aspirin. Some also showed a reduced risk associated with non-steroidal anti-inflammatory drugs other than aspirin, but the relationship was generally not as strong as with aspirin. There was no association with use of paracetamol. The possibility that the association might be accounted for by earlier detection of adenomas because of aspirin-induced bleeding was considered, but seems unlikely as no decreased risk of colon cancer was shown in patients treated with anticoagulants [27].

Numerous prospective studies have also found aspirin use to be associated with a 26–48% lower risk of developing colorectal cancer [28–30] and colorectal adenomas [29,31]. In general the association was stronger with more consistent aspirin use and persisted after adjustment for multiple risk factors. The Nurses' Health Study suggested that benefit may not be evident until after at least a decade of regular aspirin consumption [30].

The American Cancer Society Study investigated mortality from colorectal cancer in over 650 000 subjects. It found that the risk of death from both colonic and rectal tumours was significantly reduced in subjects reporting prior aspirin use measured by a brief self-report questionnaire. The association was strongest among those using aspirin 16 or more times per month for at least 10 years, with about a 64% risk reduction in fatal colon cancer and 61% reduction in fatal rectal cancer [32,33].

Patients with rheumatoid arthritis who had been treated with non-steroidal anti-inflammatory drugs including aspirin were also found to have about a 30% lower risk of colorectal cancer, especially in women [34–36], and it has been reported that sulphasalazine, which releases salicylate in the lower bowel, may reduce the risk of colorectal cancer among patients with ulcerative colitis [37].

Only in one cohort study was a positive association demonstrated between aspirin use at baseline and subsequent incidence of colorectal cancer, with a relative increase in risk of about 50% [38]. However, this study also appeared to show an increased risk of ischaemic heart disease in daily aspirin users and the study population of elderly residents of a Californian retirement community were probably highly atypical.

INTERVENTION TRIALS

The US Physicians' Health Study is the only randomized controlled trial to examine the possible primary preventive effects of aspirin against the development of colorectal cancer. After nearly five years of follow-up, subjects taking aspirin (325 mg on alternate days) had a non-significant 15% increase in risk of invasive colorectal cancer and non-significant 14% lower risk of *in situ* cancer or adenomatous or hyperplastic polyps combined compared to those treated with placebo [39]. However, the negative result may reflect inadequate dosage or duration of aspirin use rather than lack of efficacy and the results were not based on routine sigmoidoscopy or colonoscopy before and during the trial. A small randomized controlled trial of aspirin (1200 mg daily) as adjuvant therapy in 66 patients with established colorectal cancer suggested no benefit in terms of survival, but lacked any degree of statistical power [40].

IMPLICATIONS FOR CLINICAL PRACTICE

Together, the experimental and epidemiological evidence strongly suggests that aspirin can favourably influence one or more of the stages of development of colorectal cancer. There are, however, many unanswered questions about the required dose, frequency and duration of treatment which will be required for prevention to be effective. Randomized controlled trials are in progress [41] and should provide more definitive answers to guide clinical practice in the future.

In the meantime, some authorities have recommended that people at risk for colorectal cancer – those with inflammatory bowel disease; breast, ovarian or endometrial cancer; a previous adenoma or large bowel cancer; those with a family history of colorectal cancer or adenoma – should consider taking prophylactic aspirin, barring any contraindication [42].

The optimum dose of aspirin that effectively inhibits tumour development is not known. Animal studies have generally used very high doses and epidemiological studies have seldom provided much information about daily dosage, although some suggest that duration of use may be more important than the daily dose [29,30,33]. The age at which aspirin is started may also be important. A recent experimental study found that low doses of aspirin were as effective as higher doses in suppressing concentrations of colorectal mucosal prostaglandins and the authors recommended use of a single, daily dose of 80 mg in future clinical studies [43].

The balance of risk to benefit will differ according to the background risk of colorectal cancer, cardiovascular disease and serious gastrointestinal and intracranial bleeding. The newer selective COX-2 inhibitors may offer theoretical advantages over aspirin in cancer prevention, but there is very little clinical experience of their use and their much greater cost is likely to prohibit their widespread use.

Prevention of other cancers

There is limited evidence that aspirin may protect against the development of some cancers other than colorectal disease. In the large American Cancer Society Study, aspirin use was inversely related to fatal stomach and oesophagus cancer [33]. Rheumatoid arthritis patients were also found to have a lower risk of stomach cancer in linkage studies, but an increased risk of lymphoma [34,35]. A lower incidence of oesophageal cancer was also reported in the National Health and Nutrition Examination Survey (NHANES 1) [44].

Studies of human gastric adenocarcinoma tissues have been found to contain significantly raised levels of COX-2 but not COX-1 messenger RNA [45]. It has also been suggested that, in the presence of other risk factors, prostaglandin E2 may exert carcinogenic effects in the oesophagus [46]. Both findings would support a role for inhibitors of prostaglandin synthesis in preventing gastro-oesophageal cancer.

Recent experimental research also indicates that aspirin may reduce the growth and multiplication of lung adenomas and carcinomas [47]. In studies of mice, aspirin was shown to inhibit multiplication of induced lung tumours by 60% [48] and in cultured human non-small cell lung cancer cells, prostaglandin levels and cell proliferation were also significantly reduced [49].

There is some suggestion from a recent case control study that aspirin may also have some preventive potential against development of breast cancer. In a case control study breast cancer risk declined with increasing exposure to non-steroidal anti-inflammatory drugs including aspirin and the greatest reduction occurred at the highest level of use [50]. However, analysis of data from the Nurses' Health Study found that regular aspirin use at baseline and subsequently was unrelated to breast cancer incidence during the succeeding 12-year period [51].

Prevention of cataract

Cataract is one of the commonest causes of significant visual impairment and blindness in the world. It is strongly related to ageing, with some degree of lens opacity having developed in two-thirds of people aged over 50 years and in nearly 100% of those over 70 years [52]. The incidence is especially high in diabetic patients. Although modern cataract surgery is very effective, the high prevalence of the condition means that associated costs are considerable. It has been estimated that if the need for surgery could be postponed by 10 years, the number of operations required would be almost halved, producing an annual saving of over a thousand million dollars in the USA alone [53]. In the Third World, where cataracts are especially prevalent, surgical treatment is still not sufficiently available and a simple protective agent would be especially valuable.

EPIDEMIOLOGICAL STUDIES

A possible association between use of aspirin and a reduced prevalence of cataract was first observed in rheumatoid arthritis patients on long-term anti-inflammatory therapy [54,55]. In patients treated with aspirin, cataracts developed on average 10 years later and were less severe than in non-aspirin treated patients. Some subsequent epidemiological studies have given further support to a possible protective effect of aspirin [56–58], whereas others found no association [38,59–61]. The contradictory findings were attributed to differences in dosage of aspirin, duration of therapy, recall bias and patient characteristics.

Most of these criticisms were addressed in a report from the Nurses' Health Study [62]. This prospectively investigated aspirin use and cataract extraction during 434 680 person-years of follow-up and found no association after accounting for other cataract risk factors, nor when assessed by age or when allowance was made for drug dosage. Indeed, among women who consumed seven or more aspirin tablets per week for 20 or more years, there was, if anything, a nonsignificantly increased risk of cataract.

EXPERIMENTAL STUDIES AND POSSIBLE MECHANISMS

Following the early epidemiological studies, the potential of aspirin for preventing development of cataracts was further investigated in various animal models. A significant protective effect was demonstrated in rats with cataracts induced by streptozocine [63,64] or galactose [65] and in rabbits with naphthalene-induced cataracts [65].

In vitro studies suggested some possible mechanisms to explain aspirin's protective action.

- Carbamylation of the lysyl and cysteinyl residues of soluble lens proteins by cyanate may contribute to cataract formation, by lowering the protein charge and leading to conformational changes which allow disulphide bonding. It is suggested that changes in cyanate binding to lens crystallin can be reduced by aspirin, which irreversibly acetylates the cysteinyl residues and protects the lens from damage [66–68]. However, further experimental studies have indicated that at therapeutic concentrations, aspirin may not effectively inhibit cyanate binding [69].
- Non-enzymatic glycation of lens proteins by glucosamine is thought to be a possible mechanism in diabetic cataract formation, the glycation endproducts that are formed leading to opacification of the lens by disrupting the short-range order between the proteins. Protein glycation has been shown to be reduced by aspirin in both *in vitro* and *in vivo* studies [63,64], although some later studies have been less supportive [70].
- Sorbitol is toxic to lens protein and aldose reductase inhibitors and other sorbitol-lowering agents have protective effects *in vitro* and delay the development of cataract in galactose-fed rats. Aspirin inhibits aldose reductase as well as having hypoglycaemic properties and so may reduce the formation of sorbitol and subsequent lens damage [71,72].
- The finding that arachidonic acid metabolites may be demonstrated in the lens has raised the possibility that inhibition of prostaglandin synthesis by aspirin may be relevant to its protective action [73]. However, indomethacin, which is a powerful prostaglandin inhibitor, has no effect on cataract formation [74].
- Free radical activity may play a role in cataract formation, causing oxidation and precipitation of lens proteins. Aspirin has been shown to have antioxidant activity, as demonstrated by substantial reducing and/or metal-chelating activity [75]. It also protects lens protein from binding with malondialdehyde, a product of lipid peroxidation [76].

CLINICAL TRIALS

There have been several randomized controlled trials of medium to high-dose aspirin which have looked for a possible protective effect on the development of cataract, although in none was this the primary goal. The British Doctors' Trial [77] found no difference in self-reported cataract between subjects who had taken aspirin 500 mg daily for five years and controls. Similarly, the UK TIA trial showed no divergence of cataract prevalence according to length of time patients had been taking aspirin (300/1200 mg daily) or placebo [78]. The Early Treatment Diabetic Retinopathy Study [79] studied patients with non-proliferative or early proliferative diabetic retinopathy and found no evidence that five years' treatment with aspirin (650 mg daily) reduced the risk of cataract development or extraction in diabetic subjects.

Only in the US Physicians' Study did there appear to be a possible protective effect of aspirin (325 mg on alternate days) for development and extraction of cataract after an average of five years of treatment and this was restricted to the younger subjects aged 40–59 years [53]. The authors concluded that the results tended to exclude any large benefit of aspirin, whilst not excluding a possible small-to-moderate benefit on the need for extraction of age-related cataract.

IMPLICATIONS FOR CLINICAL PRACTICE

At present there is no convincing evidence to prove or disprove a potentially useful effect of aspirin in cataract prevention. In the absence of any other serious contenders, it is certainly deserving of further evaluation in large clinical trials, but cannot be recommended for routine clinical use [80,81].

Pregnancy

For at least the first part of the century, aspirin was the drug most frequently taken by women during pregnancy, to treat headaches and other general aches and pains. In the late 1960s, attitudes changed and most authorities began to advise against the use of aspirin, partly as a result of the general tendency to limit all but essential medications but also because of specific fears that it might cause teratogenic effects, increase risk of maternal and foetal haemorrhage and lead to premature closure of the ductus arteriosus [82]. In 1979 a report was published that suggested that pregnant women taking aspirin were less likely to develop pre-eclampsia than those who had not taken it [83,84]. Further reports followed and by the early 1990s the majority of obstetricians were prescribing low-dose aspirin therapy in the belief that it might prevent development or reduce severity of pre-eclampsia and foetal growth retardation in high-risk women. However, the negative outcome of the large Collaborative Low-dose Aspirin Study in Pregnancy (CLASP) [85] has now cast doubt on its efficacy in routine practice [86].

PRE-ECLAMPSIA

Hypertension complicating pregnancy and pre-eclampsia are associated with increased foetal morbidity and mortality. An increase in the ratio of thromboxane to prostacyclin has been implicated in the pathogenesis of the condition, leading to reduced perfusion, increased vascular resistance, thrombotic phenomena and placental ischaemia with resultant impaired foetal growth and the clinical features of pre-eclampsia [87]. The antiplatelet effect of low-dose aspirin was suggested as a method of reducing early platelet aggregability and so placental ischaemia [88].

The first clinical study was published by Michel Beaufils and colleagues in 1985 and claimed significant protection against the development of pre-eclampsia in high-risk women given aspirin (150 mg daily) plus dipyridamole from the start of the second trimester [89]. The results of subsequent studies [90–93] appeared to support this finding and a meta-analysis of completed trials in 1991 suggested a lower incidence of proteinuric hypertension and of major complications such as intrauterine growth retardation and foetal death in treated patients [94]. As might be expected, the effect appeared to be most significant in patients at greatest risk.

However, the much larger CLASP study of nearly 10 000 women [85] failed to show any statistically significant benefit of aspirin (60 mg daily) over placebo, except for a small reduction (from 19% to 17%) in preterm delivery. Neither placental haemorrhage nor maternal bleeding during epidural anaesthesia was more likely on aspirin treatment, but there was a greater need for blood transfusion after delivery. The negative findings of this study may have been influenced by the relatively low-risk patients who took part. Another trial involving more than 3000 healthy, nulliparous pregnant women also found no benefit of aspirin and a significant increase in risk of antepartum haemorrhage [95]. The Italian study of aspirin in pregnancy was similarly negative [96].

A meta-analysis of 13 published randomized clinical trials conducted between 1985 and 1994 of prophylactic low-dose aspirin in pregnancy showed a significant 18% reduction in intrauterine growth retardation and a non-significant 16% reduction in perinatal mortality. Subgroup analysis suggested that although lower doses were protective, greatest reductions were seen at doses of aspirin between 100 and 150 mg daily and in women starting treatment before the 17th week of gestation. The authors concluded, however, that low-dose aspirin should not be used routinely in pregnant women until those most likely to benefit from aspirin treatment had been clearly identified [97]. At present, the evidence suggests that these are most likely to be women who have a personal or strong family history of early pre-eclampsia and who can commence aspirin therapy before 20 weeks of pregnancy. The safety of aspirin in this indication would seem well established.

MISCARRIAGE

Recurrent foetal loss has been well described in women with antiphospholipid antibodies and about 15% of women with a history of recurrent miscarriage have positive results for antiphospholipid antibodies on screening [98,99]. Thrombosis of the placental vasculature is assumed to be the underlying mechanism [100] and so antithrombotic regimens would seem to be logical treatment.

Two recently published trials have compared the use of low-dose aspirin plus heparin with low-dose aspirin alone [101,102] in women with a history of recurrent miscarriage in association with phospholipid antibodies. They concluded that the poor obstetric outlook in these patients may be improved with low-dose aspirin, but it is further and significantly improved with combination antithrombotic therapy of low-dose aspirin and low-dose heparin. However, a positive antiphospholipid antibody test result in a pregnant woman with no other signs or symptoms of the antiphospholipid antibody syndrome is probably not an indication for low-dose aspirin as the risk of obstetric complications is very low [103].

The combination of aspirin and prednisone was not effective in promoting live birth in a recent placebo-controlled trial in pregnant women with antinuclear, anti-DNA or antilymphocyte autoantibodies or antibodies for the lupus anticoagulant and more infants were born prematurely in the active treatment group [104].

Prevention of cognitive decline and dementia

Decline in cognitive function in old age is common and of growing clinical and public health concern. There is now a large body of experimental and epidemiological evidence suggesting that much of this age-related cognitive decline and dementia is associated with atheromatous and cardiovascular disease and the traditional division between vascular and degenerative dementia is increasingly under threat.

The large population-based Rotterdam Study found that people with evidence of peripheral arterial disease, carotid artery plaques detected by ultrasound or electrocardiographic evidence of past myocardial infarction showed more impaired cognitive test scores compared to subjects not so exposed [105] and the Medical Research Council Elderly Hypertension

Trial has also reported poorer cognitive function associated with evidence of ischaemic heart disease in elderly hypertensive patients [106]. Stroke is well recognized as a major risk factor for cognitive impairment and vascular dementia is typically the result of multiple cerebral infarctions or a single strategic infarct [107]. Vascular lesions may also contribute to the clinical manifestation of Alzheimer's disease, due to summation of subclinical vascular and Alzheimer lesions. Recent reports of an excessive microvascular pathology [108] and blood–brain barrier dysfunction [109] associated with Alzheimer's disease might imply a more direct role of vascular abnormalities in the pathogenesis of the illness [110]. All these findings suggest that low-dose aspirin might have an important role to play in preventing progression of cerebrovascular pathology and thus preserve cognitive function.

Further evidence suggesting a protective effect of aspirin therapy comes from epidemiological studies reporting an inverse relationship between use of aspirin and other non-steroidal anti-inflammatory drugs and the risk of developing cognitive impairment and Alzheimer's disease [111,112]. Both the initial development of symptoms as well as their subsequent progression appear to be prevented or delayed [113]. Other studies, however, have shown no association between aspirin use and cognitive test scores or decline [114,115]. It is suggested that any possible beneficial effect may be attributed to control of the inflammatory response in the brain or to a free radical scavenging action, as well as to antithrombotic mechanisms. Surprisingly, the Canadian Study of Health and Aging found that aspirin use was positively associated with risk of vascular dementia (odds ratio 3.1), but the finding was interpreted as use of aspirin being a predictor of survival rather than a risk factor [116].

A recent report from the Thrombosis Prevention Trial gave details of cognitive test score results in 405 older men at risk of cardiovascular disease who had been participating in a randomized factorial trial of low-dose aspirin (75 mg daily) and/or low intensity oral anticoagulation or placebo for at least five years. Verbal fluency and mental flexibility were significantly better in subjects taking antithrombotic medication than in subjects taking placebo and aspirin was considered to have contributed more than warfarin to any beneficial effect [117]. A randomized controlled trial of the effect on cognitive function of aspirin (100 mg daily) versus placebo is in progress in men in the Caerphilly Collaborative Heart Disease Study [118].

There has only been one published randomized controlled trial of aspirin (325 mg daily) in patients with established vascular dementia. Highly significant improvements were demonstrated for cerebral perfusion values and cognitive performance scores among aspirin-treated patients compared to untreated controls at each of three annual follow-up evaluations. Both men and women appeared to benefit from aspirin therapy and their quality of life and independence were said to have improved, which was not apparent in the control group. The authors concluded that daily aspirin appears to improve or stabilize declines in cerebral perfusion and cognition among patients with multi-infarct dementia. Whilst methodologically this trial was seriously flawed, the reported results are impressive and would certainly justify further large-scale studies [119].

Despite lack of definitive evidence from randomized controlled trials, low-dose aspirin is widely accepted as appropriate secondary preventive therapy in patients with probable vascular dementia, to reduce the risk of recurrent ischaemic episodes. It is premature to promote aspirin in antithrombotic or anti-inflammatory doses for prevention or treatment of age-related cognitive impairment or Alzheimer's disease. The results of ongoing randomized controlled trials should decide if it is effective, who is most likely to benefit, what is the optimum dose and when it should be started.

Other possible therapeutic uses

LEG ULCERS

A single randomized controlled trial of low-dose aspirin (300 mg) versus placebo in 20 patients with chronic venous leg ulcers suggested that after four months treatment healing was greater in patients taking active treatment [120]. The possible mechanism of action was unclear [121] and the validity of the study has been questioned [122].

ALLERGIC RHINITIS AND CONJUNCTIVITIS

Because of the involvement of prostaglandins in the production of symptoms of allergic rhinitis and conjunctivitis, aspirin and other non-steroidal anti-inflammatory drugs have been investigated as potential treatments. A small randomized controlled trial compared topical use of aspirin eye drops (1% solution, applied as one drop in each eye, four times daily for 14 days) and placebo in patients with seasonal allergic conjunctivitis. Significant benefit was claimed and no serious side effects were observed [123]. Aspirin has also been suggested as a prophylactic agent against food hypersensitivity reactions.

ACQUIRED IMMUNE DEFICIENCY SYNDROME (AIDS)

The finding that salicylate and aspirin, but not ibuprofen or paracetamol, block the formation of nuclear factor-kappa B raises the possibility of use of aspirin as an adjunctive treatment in acquired immune deficiency syndrome (AIDS) [16]. There are some anecdotal reports of aspirin increasing T4 cell counts in HIV-positive subjects, but no published controlled clinical trials.

PERIODONTAL DISEASE

It has been suggested that long-term low-dose aspirin may improve circulation to the gums, thereby arresting periodontal disease. This is supported by the results of one small study in rats, where aspirin administration appeared to suppress root resorption induced by mechanical injury to the periodontal soft tissues [124]. There is as yet no evidence of any significant effects in humans.

MOOD-MODULATING EFFECTS

The possibility that aspirin may have a psychoactive effect which is partly responsible for its beneficial effects in ischaemic heart disease has been suggested by a small study in men undergoing coronary angiography [125]. Aspirin use was associated with less depression and anxiety or worry, as reported by the patient and as perceived by a significant other, and the association could not be accounted for by results of cardiological investigations or other clinical and demographic variables. The authors concluded that, although only correlational in nature, the results raise the question of whether aspirin may have a beneficial mood-modulating effect.

OSTEOPOROSIS

Aspirin may inhibit bone loss and preserve bone mineral density *in vitro* and in animal models of osteoporosis. A study in postmenopausal women found that daily use of aspirin was associated with a 2–6% increase in bone mineral density of the hip and spine over the four years of follow-up compared to no use of aspirin, but there was no difference in fracture risk [126].

MORTALITY AFTER HIP FRACTURE

A study of determinants of short-term mortality in elderly patients with hip fracture found that postoperative use of aspirin was associated with a significantly reduced risk of death. Cardiovascular events were the presumed cause in 63% of in-hospital deaths and the authors suggested that aspirin might therefore have considerable potential to reduce mortality in this population [127].

Conclusions

There is accumulating evidence that aspirin may have a role in several new therapeutic areas, the strongest support being for use to prevent gastrointestinal and possibly other cancers, cataract, cognitive decline and certain high-risk pregnancies. There may be difficulties in the future with organizing the required randomized controlled trials to prove beyond doubt any benefit from aspirin in these new indications. For adequate statistical power to provide unequivocal answers, most trials would require the recruitment of very large numbers of subjects representative of the at-risk population and followed up for at least several years. However, this may become increasingly difficult as any patients at risk of cardiovascular disease could not ethically be maintained on placebo, many other potential subjects will already be taking aspirin and others may be unwilling to be randomly allocated or comply with the requirements of the control group, given the increasing lay knowledge of aspirin's preventive potential. It would seem to be a matter of urgency that the necessary trials are commenced.

KEYPOINTS

- Epidemiological and experimental evidence consistently shows that aspirin may inhibit the occurrence or growth of gastrointestinal tumours, especially of the large bowel. Clinical trials are in progress.
- Less consistent evidence suggests a role for aspirin in prevention of cataract and of cognitive decline and dementia.
- The use of low-dose aspirin to prevent and treat pre-eclampsia and foetal growth retardation may be justified in subjects at high risk. It has also proved valuable in preventing recurrent miscarriage associated with antiphospholipid antibodies.

REFERENCES

1. Craven PA, DeRubertis FR. Effects of aspirin on 1,2-dimethylhydrazine-induced colonic carcinogenesis. Carcinogenesis 1992; 13: 541–546.
2. DeRubertis FR, Craven PA. Early alterations in rat colonic mucosal cyclic nucleotide metabolism and protein kinase activity induced by 1,2-dimethylhydrazine. Cancer Res 1980; 40: 4589–4598.
3. Davis AE, Patterson F. Aspirin reduces the incidence of colonic carcinoma in the dimethylhydrazine rat animal model. Aust NZ J Med 1994; 24: 301–303.

4. Reddy BS, Rao CV, Rivenson A, Kelloff G. Inhibitory effect of aspirin on azoxymethane-induced colon carcinogenesis in F344 rats. Carcinogenesis 1993; 14: 1493–1497.
5. Wargovich MJ, Chen CD, Harris C *et al.* Inhibition of aberrant crypt growth by non-steroidal antiinflammatory agents and differentiation agents in the rat colon. Int J Cancer 1995; 60: 515–519.
6. Craven PA, Thornburg K, DeRubertis FR. Sustained increase in the proliferation of rat colonic mucosa during chronic treatment with aspirin. Gastroenterology 1988; 94: 567–575.
7. Gustafson-Svard C, Lilja I, Hallbook O, Sjodahl R. Cyclo-oxygenase and colon cancer: clues to the aspirin effect? Ann Med 1997; 29: 247–252.
8. Bennett A, Del Tacca M, Stamford IF, ZebroT. Prostaglandins from tumours of human large bowel. Br J Cancer 1977; 35: 881–884.
9. Narisawa T, Kusaka H, Yamazaki Y *et al.* Relationship between blood plasma prostaglandin E2 and liver and lung metastases in colorectal cancer. Dis Colon Rectum 1990; 33: 840–845.
10. Marnett LJ. Aspirin and the role of prostaglandins in cancer. Cancer Res 1992; 52: 5575–5589.
11. Rigas B, Goldman IS, Levine L. Altered eicosanoid levels in human colon cancer. J Lab Clin Med 1993; 122: 518–523.
12. Eberhart CE, Coffey RJ, Radhika A *et al.* Up-regulation of cyclooxygenase-2 gene expression in human colorectal adenomas and adenocarcinomas. Gastroenterology 1994; 107: 1183–1188.
13. DuBois RN, Tsujii M, Bishop P *et al.* Cloning and characterisation of a growth factor inducible cyclooxygenase gene from rat intestine epithelial cells. Am J Physiol 1994; 266: 822–829.
14. Sheng H, Shao J, Kirkland SC *et al.* Inhibition of human colon cancer cell growth by selective inhibition of cyclooxygenase-2. J Clin Invest 1997; 99: 2254–2259.
15. Setty BNY, Graever JE, Stuart MJ. The mitogenic effect of 15- and 12-hydroxyeicosatetraenoic acid on endothelial cells may be mediated via diacylglycerol kinase inhibition. J Biol Chem 1987; 262: 17613–17622.
16. Kopp E, Ghosh S. Inhibition of NF-kappa B by sodium salicylate and aspirin. Science 1994; 265: 956–959.
17. Dong Z, Huang C, Brown RE, Ma WY. Inhibition of activator protein 1 activity and neoplastic transformation by aspirin. J Biol Chem 1997; 272: 9962–9970.
18. Finley G, Melhem M. Cytokine expression in colorectal adenocarcinoma and mucosa. Proc Am Assoc Cancer Res 1996; 37: A1090.
19. Sciff Sj, Qiao L, Tsai L, Rigas B. Sulindac sulfide, an aspirin-like compound inhibits proliferation, causes cell cycle quiescence, and induces apoptosis in HT-29 colon adenocarcinoma cells. J Clin Invest 1995; 96: 491–503.
20. Morgan G. Aspirin chemoprevention of colorectal and oesophageal cancers. An overview of the literature and homeopathic explanation. Eur J Cancer Prev 1996; 5: 439–443.

21. Kune GA, Kune S, Watson LF. Colorectal cancer risk, chronic illnesses, operations, and medications: case control results from the Melbourne Colorectal Cancer Study. Cancer Res 1988; 48: 4399–4404.
22. Rosenberg L, Palmer JR, Zauber AG *et al.* A hypothesis: nonsteroidal anti-inflammatory drugs reduce the incidence of large-bowel cancer. J Natl Cancer Inst 1991; 83: 355–358.
23. Peleg II, Maibach HT, Brown SH, Wilcox CM. Aspirin and nonsteroidal antiinflammatory drug use and the risk of subsequent colorectal cancer. Arch Intern Med 1994; 154: 393–399.
24. Muscat JE, Stellman SD, Wynder EL. Nonsteroidal antiinflammatory drugs and colorectal cancer. Cancer 1994; 74: 1847–1854.
25. Suh O, Mettlin C, Petrelli NJ. Aspirin use, cancer, and polyps of the large bowel. Cancer 1993; 72: 1171–1177.
26. Logan RFE, Little J, Hawtin PG, Hardcastle JD. Effect of aspirin and non-steroidal antiinflammatory drugs on colorectal adenomas: case-control-study of subjects participating in the Nottingham faecal occult blood screening programme. Br Med J 1993; 307: 285–289.
27. Muller AD, Sonneberg A, Wasserman IH. Diseases preceding colon cancer: a case-control study among veterans. Dig Dis Sci 1994; 39: 2480–2484.
28. Schreinemachers DM, Everson RB. Aspirin use and lung, colon, and breast cancer incidence in a prospective study. Epidemiology 1994; 5: 138–146.
29. Giovannucci E, Rimm EB, Stampfer MJ *et al.* Aspirin use and the risk for colorectal cancer and adenoma in male health professionals. Ann Intern Med 1994; 121: 241–246.
30. Giovannucci E, Egan KM, Hunter DJ *et al.* Aspirin and the risk of colorectal cancer in women. N Engl J Med 1995; 333: 609–614.
31. Greenberg ER, Baron JA, Freeman DH Jr, Mandel JS, Haile R. Reduced risk of large bowel adenomas among aspirin users. J Natl Cancer Inst 1993; 85: 912–916.
32. Thun MJ, Namboodiri MM, Heath CW Jr. Aspirin use and reduced risk of fatal colon cancer. N Engl J Med 1991; 325: 1593–1596.
33. Thun MJ, Namboodiri MM, Calle EE, Flanders WD, Heath CW Jr. Aspirin use and risk of fatal cancer. Cancer Res 1993; 53: 1322–1327.
34. Gridley G, McLoughlin JK, Ekbom A *et al.* Incidence of cancer among patients with rheumatoid arthritis. J Natl Cancer Inst 1993; 85: 307–311.
35. Isomaki HA, Hakulinen T, Joutsenlahti U. Excess risk of lymphomas, leukaemia and myeloma in patients with rheumatoid arthritis. J Chron Dis 1978; 31: 691–696.
36. Laakso M, Mutru O, Isomaki H, Koota K. Cancer mortality in patients with rheumatoid arthritis. J Rheumatol 1986; 13: 522–526.
37. Pinczowski D, Ekbom A, Baron J, Yuen J, Adami H-O. Risk factors for colorectal cancer in patients with ulcerative colitis: a case-control study. Gastroenterology 1994; 107: 117–120.
38. Paganini-Hill A, Chao A, Ross RK, Henderson BE. Aspirin use and chronic diseases: a cohort study of the elderly. Br Med J 1989; 299: 1247–1250.

39. Gann PH, Manson JE, Glynn RJ, Burling JE, Hennekens CH. Low-dose aspirin and incidence of colorectal tumors in a randomized trial. J Natl Cancer Inst 1993; 85: 1220–1224.
40. Lipton A, Scialla S, Harvey H *et al.* Adjuvant antiplatelet therapy with aspirin in colorectal cancer. J Med 1982; 23: 419–429.
41. Burn J, Chapman PD, Mathers J *et al.* The protocol for the European double blind trial of aspirin and resistant starch in familial adenomatous polyposis: the CAPP Study. Conc Action Polyposis Prev 1995; 31A: 1385–1386.
42. Marcus AJ. Aspirin as prophylaxis against colorectal cancer. N Engl J Med 1995; 333: 636–638.
43. Ruffin MT 4th, Krishnan K, Rock CL *et al.* Suppression of human colorectal mucosal prostaglandins: determining the lowest effective aspirin dose. J Natl Cancer Inst 1997; 89: 1152–1160.
44. Funkhouser EM, Sharp GB. Aspirin and reduced risk of oesophageal cancer. Cancer 1995; 76: 1116–1119.
45. Ristimaki A, Honkanen N, Jankala H, Sipponen P, Harkonen M. Expression of cyclooxygenase-2 in human gastric carcinoma. Cancer Res 1997; 57: 1276–1280.
46. Morgan G. Deleterious effects of prostaglandin E2 in oesophageal carcinogenesis. Med Hypotheses 1997; 48: 177–181.
47. Anonymous. Aspirin prevention update: new data on lung and colon cancers. J Natl Cancer Inst 1995; 87: 567–569.
48. Duperron C, Castonguay A. Chemopreventive efficacies of aspirin and sulindac against lung tumorigenesis in A/J mice. Carcinogenesis 1997; 18: 1001–1006.
49. Avis IM, Jett M, Boyle T *et al.* Growth control of lung cancer by interruption of 5r lipoxygenase-mediated growth factor signaling. J Clin Invest 1996; 97: 806–813.
50. Harris RE, Namboodri KK, Farra WB. Nonsteroidal antiinflammatory drugs and breast cancer. Epidemiology 1996; 7: 203–205.
51. Egan KM, Stampfer MJ, Giovannucci E, Rosner BA, Colditz GA. Prospective study of regular aspirin use and the risk of breast cancer. J Natl Cancer Inst 1996; 88: 988–993.
52. Harding J. Cataract: Biochemistry, Epidemiology and Pharmacology. London: Chapman & Hall, 1991.
53. Seddon JM, Christen WG, Manson JE *et al.* Low-dose aspirin and risk of cataract in a randomised trial of US physicians. Arch Ophthalmol 1991; 109: 252–255.
54. Cotlier E. Rheumatoid arthritis and cataract surgery. Int Ophthalmol 1980; 2: 127–129.
55. Cotlier E, Sharma YR. Aspirin and senile cataracts in rheumatoid arthritis. Lancet 1981; i: 338–339.

56. Van Heyningen R, Harding JJ. Do aspirin-like analgesics protect against cataract? A case-control study. Lancet 1986; i: 1111–1113.
57. Harding JJ, Egerton M, Harding RS. Protection against cataract by aspirin, paracetamol and ibuprofen. Acta Ophthalmol 1989; 67: 518–524.
58. Mohan M, Sperduto RD, Angra SK *et al.* India–US case-control study of age-related cataracts. Arch Ophthalmol 1989; 107: 670–676.
59. Kewitz H, Nitz M, Gaus V. Aspirin and cataract. Lancet 1986; i: 689.
60. West SK, Munoz BE, Newland HS, Emmett EA, Taylor HR. Lack of evidence for aspirin use and prevention of cataracts. Arch Ophthalmol 1987; 105: 1229–1231.
61. Klein BE, Klein R, Moss SE. Is aspirin use associated with lower rates of cataracts in diabetic individuals? Diabetes Care 1987; 10: 495–499.
62. Hankinson SE, Seddon JM, Colditz GA *et al.* A prospective study of aspirin use and cataract extraction in women. Arch Ophthalmol 1993; 111: 503–508.
63. Swamy MS, Abraham EC. Inhibition of lens crystallin glycation and high molecular weight aggregate formation by aspirin in vitro and in vivo. Invest Ophthalmol Vis Sci 1989; 30: 1120–1126.
64. Blakytny R, Harding JJ. Prevention of cataract in diabetic rats by aspirin, paracetamol (acetaminophen) and ibuprofen. Exp Eye Res 1992; 54: 509–518.
65. Gupta PP, Pandey DN, Pandey DJ *et al.* Aspirin in experimental cataractogenesis. Indian J Med Res 1984; 80: 703–707.
66. Rao GN, Lardis MP, Cotlier E. Acetylation of lens crystallins: a possible mechanism by which aspirin could prevent cataract formation. Biochem Biophys Res Commun 1985; 128: 1125–1132.
67. Crompton M, Rixon KC, Harding JJ. Aspirin prevents carbamylation of soluble lens proteins and prevents cyanate-induced phase separation opacities in vitro: a possible mechanism by which aspirin could prevent cataract. Exp Eye Res 1985; 40: 297–311.
68. Qin W, Smith JB, Smith DL. Reaction of aspirin with cysteinyl residues of lens gamma-crystallins: a mechanism for the proposed anti-cataract effect of aspirin. Biochim Biophys Acta 1993; 1181: 103–110.
69. Hasan A, Smith JB, Qin W, Smith DL. The reaction of bovine lens alpha A-crystallin with aspirin. Exp Eye Res 1993; 57: 29–35.
70. Stevens A. The effectiveness of putative anti-cataract agents in the prevention of protein glycation. J Am Optometric Assoc 1995; 66: 744–749.
71. Kinoshita JH. Mechanisms initiating cataract formation. Invest Ophthalmol 1974; 13: 713–724.
72. Kinoshita JH, Fukushi S, Kador HN. Aldose reductase in diabetic complications of the eye. Metabolism 1979; 28: 462–469.

73. Zheng DR, Fu SCJ, Lysz TW, Leung CCK. Immunocytochemical localization of cyclooxygenase in the rat lens. Invest Ophthalmol Vis Sci 1992; 33: 178–183.
74. Harding JJ, Van Heyningen R. Aspirin-like analgesics and cataract. Lancet 1986; ii: 1044–1045.
75. Woollard AC, Wolff SP, Bascal ZA. Antioxidant characteristics of some potential anticataract agents. Studies of aspirin, paracetamol, and bendazac provide support for an oxidative component of cataract. Free Rad Biol Med 1990; 9: 299–305.
76. Riley ML, Harding JJ. The reaction of malondialdehyde with lens proteins and the protective effect of aspirin. Biochim Biophys Acta 1993; 1158: 107–112.
77. Peto R, Gray R, Collins R *et al.* Randomised trial of prophylactic daily aspirin in British male doctors. Br Med J 1988; 296: 313–316.
78. UK TIA Study Group. Does aspirin affect the rate of cataract formation? Cross-sectional results during a randomised double-blind placebo controlled trial to prevent serious vascular events. Br J Ophthalmol 1992; 76: 259–261.
79. Chew EY, Williams GA, Burton TC *et al.* Aspirin effects on the development of cataracts in patients with diabetes mellitus: Early Treatment Diabetic Retinopathy Study report 16. Arch Ophthalmol 1992; 110: 339–342.
80. Editorial. Preventing cataract. Lancet 1992; 350: 883–884.
81. Harding JJ. Pharmacological treatment strategies in age-related cataracts. Drugs Aging 1992; 2: 287–300.
82. Collins E, Turner G. Maternal effects of regular salicylate ingestion. Lancet 1975; ii: 335–357.
83. De Swiet M, Fryers G. The use of aspirin in pregnancy. J Obstet Gynaecol 1990; 10: 467–482.
84. Crandon AJ, Isherwood DM. Effect of aspirin on incidence of preeclampsia. Lancet 1979; i: 1356.
85. CLASP Collaborative Group. CLASP: a randomised trial of low-dose aspirin for the prevention and treatment of pre-eclampsia among 9364 pregnant women. Lancet 1994; 343: 619–629.
86. Beilin L. Aspirin and pre-eclampsia. Br Med J 1994; 308: 1250–1251.
87. Walsh SW. Preeclampsia: an imbalance in placental prostacyclin and thromboxane production. Am J Obstet Gynecol 1985; 152: 335–340.
88. Redman CWG, Bonnar J, Beilin L. Early platelet consumption in pre-eclampsia. Br Med J 1978; i: 467–469.
89. Beaufils M, Uzan S, Donsimoni R, Colau JC. Prevention of pre-eclampsia by early platelet therapy. Lancet 1985; i: 840–842.
90. Wallenburg HCS, Rotmans N. Prevention of recurrent idiopathic fetal growth retardation by low-dose aspirin and dipyridamole. Am J Obstet Gynecol 1987; 157: 1230–1235.

91. Benigni A, Gregorini G, Frusca T *et al.* Effect of low-dose aspirin on fetal and maternal generation of thromboxane by platelets in women at risk for pregnancy-induced hypertension. N Engl J Med 1989; 321: 357–362.
92. Schiff E, Peleg E, Goldenberg M *et al.* The use of aspirin to prevent pregnancy-induced hypertension and lower the ratio of thromboxane A2 to prostacyclin in relatively high risk pregnancies. N Engl J Med 1989; 321: 351–356.
93. McParland P, Pearce JM, Chamberlain GVP. Doppler ultrasound and aspirin in recognition and prevention of pregnancy-induced hypertension. Lancet 1990; 335: 1552–1555.
94. Imperiale TF, Petrulis AS. A meta-analysis of low-dose aspirin for the prevention of pregnancy induced hypertensive disease. JAMA 1991; 266: 260–264.
95. Sibai BM, Caritis SN, Thom E *et al.* Prevention of preeclampsia with low-dose aspirin in healthy, nulliparous pregnant women. N Engl J Med 1993; 329: 1213–1218.
96. Italian Study of Aspirin in Pregnancy. Low-dose aspirin in prevention and treatment of intrauterine growth retardation and pregnancy-induced hypertension. Lancet 1993; 341: 396–400.
97. Leitich H, Egarter C, Husslein P, Kaider A, Schemper M. A meta-analysis of low-dose aspirin for the prevention of intrauterine growth retardation. Br J Obstet Gynaecol 1997; 104: 450–459.
98. Rai RS, Clifford K, Cohen H, Regan L. High prospective fetal loss rate in untreated pregnancies of women with recurrent miscarriage and antiphospholipid antibodies. Hum Reprod 1995; 10: 3301–3304.
99. Rai RS, Regan L, Clifford K *et al.* Antiphospholipid antibodies and beta-2 glycoprotein-I in 500 women with recurrent miscarriage: results of a comprehensive screening approach. Hum Reprod 1995; 10: 101–105.
100. Out HJ, Koojiman CD, Brinse HW, Derksen RH. Histopathological findings in placentae from patients with intra-uterine fetal death and anti-phospholipid antibodies. Eur J Obstet Gynecol Reprod Biol 1991; 41: 179–186.
101. Kutteh WH. Antiphospholipid antibody-associated recurrent pregnancy loss: treatment with heparin and low-dose aspirin is superior to low dose aspirin alone. Am J Obstet Gynecol 1996; 174: 1584–1589.
102. Rai R, Cohen H, Dave M, Regan L. Randomised controlled trial of aspirin and aspirin plus heparin in pregnant women with recurrent miscarriage associated with phospholipid antibodies (or antiphospholipid antibodies). Br Med J 1997; 314: 253–257.
103. Cowchock S, Reece EA. Do low-risk pregnant women with antiphospholipid antibodies need to be treated? Organising Group of the Antiphospholipid Antibody Treatment Trial. Am J Obstet Gynecol 1997; 176: 1099–1100.
104. Laskin CA, Bombardier C, Hannah ME *et al.* Prednisone and aspirin in women with autoantibodies and unexplained recurrent fetal loss. N Engl J Med 1997; 337: 148–153.

105. Breteler MM, Van Swieten JC, Bots ML *et al.* Cerebral white matter lesions, vascular risk factors, and cognitive function in a population-based study: the Rotterdam Study. Neurology 1994; 44: 1246–1252.
106. Prince MJ. Vascular risk factors and atherosclerosis as risk factors for cognitive decline and dementia. J Psychosom Res 1995; 39: 525–530.
107. Carlson LA, Gottfries CG, Winblad B. Vascular dementia. Dementia 1994; 5: 129–214.
108. Buee L, Hof PR, Bouras C. Pathological alterations of the cerebral microvasculature in Alzheimer's disease and related disorders. Acta Neuropathol 1994; 57: 469–480.
109. Mattila KM, Pirtilla T, Blennow K *et al.* Altered blood–brain barrier function in Alzheimer's disease. Acta Neurol Scand 1994; 89: 192–198.
110. De La Torre JC. Cerebromicrovascular pathology in Alzheimer's disease compared to normal aging. Gerontology 1997; 43: 26–43.
111. McGeer PL, McGeer E, Rogers J, Sibley J. Antiinflammatory drugs and Alzheimer's disease. Lancet 1990; 335: 1037.
112. Rich JB, Rasmusson DX, Folstein MF *et al.* Nonsteroidal anti-inflammatory drugs in Alzheimer's disease. Neurology 1995; 45: 51–55.
113. Breitner JC, Gau BA, Welsh KA *et al.* Inverse association of anti-inflammatory treatments and Alzheimer's disease: initial results of a co-twin control study. Neurology 1994; 44: 227–232.
114. May FE, Moore MT, Stewart RB, Hale WE. Lack of association of non-steroidal anti-inflammatory drug use and cognitive decline in the elderly. Gerontology 1992; 38: 275–279.
115. Sturmer T, Glynn RJ, Field TS, Taylor JO, Hennekens CH. Aspirin use and cognitive function in the elderly. Am J Epidemiology 1996; 143: 683–691.
116. Lindsay J, Hebert R, Rockwood K. The Canadian Study of Health and Aging: risk factors for vascular dementia. Stroke 1997; 28: 526–530.
117. Richards M, Meade TW, Peart S, Brennan PJ, Mann AH. Is there any evidence for a protective effect of antithrombotic medication on cognitive function in men at risk of cardiovascular disease? Some preliminary findings. J Neurol Neurosurg Psychiatr 1997; 62: 269–272.
118. Elwood P, Hughes C. Aspirin and Cardiovascular Disease. Cardiff: Crystal Print, 1997.
119. Meyer JS, Rogers RL, McClintic K, Mortel KF, Lotfi J. Randomized clinical trial of daily aspirin therapy in multi-infarct dementia. A pilot study. J Am Geriatr Soc; 37: 49–55.
120. Layton AM, Ibbotson SH, Davies JA, Goodfield MJ. Randomised trial of oral aspirin for chronic venous leg ulcers. Lancet 1994; 344: 164–165.
121. Ibbotson SH, Layton AM, Davies JA, Goodfield MJ. The effect of aspirin on haemostatic activity in the treatment of chronic venous leg ulceration. Br J Dermatol 1995; 132: 422–426.
122. Ruckley CV, Prescott RJ. Treatment of chronic leg ulcers. Lancet 1994; 344: 1512–1513.

123. Ciprandi G, Buscaglia S, Tosca M, Canonica GW. Topical acetylsalicylic acid in the treatment of allergic pollinosic conjunctivitis. J Invest Allergol Clin Immunol 1992; 2: 15–18.
124. Kameyama Y, Nakane S, Maeda H, Fujita K, Takesue M, Sato E. Inhibitory effect of aspirin on root resorption induced by mechanical injury of the soft periodontal tissues in rats. J Periodont Res 1994; 29:113–117.
125. Ketterer MW, Brymer J, Rhoads K, Kraft P, Lovallo WR. Is aspirin, as used for antithrombosis, an emotion-modulating agent? J Psychosomat Res 1996; 40: 53–58.
126. Bauer DC, Orwell ES, Fox KM *et al.* Aspirin and NSAID use in older women: effect on bone mineral density and fracture risk. Study of Osteoporotic Fractures Research Group. J Bone Min Res 1996; 11: 29–35.
127. Nettleman MD, Alsip J, Schrader M, Schulte M. Predictors of mortality after hip fracture. J Gen Intern Med 1996; 11: 765–767.

Adverse effects and toxicity

Introduction

All effective drugs cause side effects in some patients. Given the widespread use of aspirin in so many different circumstances over so many years, it is perhaps not too surprising that a multiplicity of associated adverse effects have been reported. Whilst occasional use of standard doses of aspirin generally causes few problems, large doses taken over a sustained period of time are associated with a much higher incidence of adverse effects and overdose can lead to serious toxicity. The gastrointestinal side effects and hypersensitivity reactions are most frequently reported in clinical use, but serious problems are unusual unless aspirin is taken improperly. In infants and young children, aspirin has been implicated in the development of Reye's syndrome and this has led to severe restrictions in its use to treat children under the age of 12 years.

Gastrointestinal side effects

The commonest reported adverse effects associated with aspirin treatment are gastrointestinal disturbances of dyspepsia, nausea and vomiting, heartburn or burning upper abdominal discomfort. These were reported by 2.1% of hospitalized patients given aspirin acutely [1] and are reported at some time by up to 25% of patients in long-term studies, a rate similar to that in patients given placebo [2]. More serious gastrointestinal side effects include peptic ulceration, gastrointestinal haemorrhage and erosive gastritis, but these are much less common [3–6].

CLINICAL PRESENTATION

Unfortunately the absence of symptoms does not predict absence of significant gastric damage. Nor does dyspepsia necessarily relate to intestinal blood loss or endoscopic appearance of the gastric mucosa. Elderly patients in particular often do not report any relevant symptoms before they present with a serious aspirin-related gastrointestinal bleed [7].

Typically, therefore, gastric bleeding is painless and may go unnoticed until the patient presents with iron deficiency anaemia or an acute haemorrhage. An anti-inflammatory dose of aspirin results in average faecal blood loss of about 3–8 ml per day, compared to about 0.6 ml per day in untreated control subjects [8]. Lower doses of aspirin, as currently used for secondary prevention, cause only a slight, clinically insignificant increase in faecal blood loss, insufficient to result in positive faecal occult blood tests [9].

Bleeding is greatest with slowly dissolving preparations which deposit as particles in the mucosal folds of the stomach [10]. Gastroscopic examination will reveal signs ranging from gastric erythema and small pinpoint haemorrhages to erosive gastritis and sharply demarcated areas of ulceration with frank haemorrhage [11]. Ulcers are most commonly situated in the antral and prepyloric areas [12]. Aspirin therapy has also been linked with the development of oesophagitis [13], both due to pill entrapment in the oesophagus inducing local irritation and by exacerbation of gastro-oesophageal reflux [14].

RISK OF GI EFFECTS IN THERAPEUTIC USE

The absolute risk of significant gastrointestinal side effects associated with aspirin therapy is low (Table 7.1). Based on observational studies in patients taking analgesic or anti-inflammatory doses, the incidence of significant dyspepsia, nausea and vomiting was estimated to be between 2% and 6% [15]. The Boston Collaborative Drug Surveillance Program was the first to show conclusively that heavy regular use of aspirin was associated with major upper gastrointestinal haemorrhage and uncomplicated gastric ulcer [16]. The incidence of severe haemorrhage or gastric ulceration was calculated to be about 10–15 per 100 000 habitual aspirin users [15].

Table 7.1
The attributable disease rate (a measure of the absolute event rate specifically due to aspirin) in patients taking aspirin in primary and secondary prevention trials

Symptom	Percentage increase in rate	Attributable rate (events/1000 person-years)
Haematemesis	40–90	0.2–1.0
Melaena	50–90	2.1–7.0
All GI bleeds	60–180	2.5–7.7
Peptic ulcer	20–30	0.5–3.3
Hospitalized for peptic ulcer	310	3.5
Cerebral haemorrhage	20–50	<0.1

Compared with 11 other non-steroidal anti-inflammatory drugs, a meta-analysis suggested that aspirin had an intermediate risk for serious gastro-intestinal complications requiring hospitalization, with a risk relative to ibuprofen of 1.6 and a ranking of fifth safest overall [17].

A recent overview has more systematically examined the gastrointestinal toxicity of aspirin in 21 placebo-controlled trials, all but one of secondary prevention [18]. These involved 40 000 patients with 75 000 person-years of aspirin exposure. All but four of the trials were using what would now be considered relatively high doses of aspirin, between 900 mg and 1500 mg daily. The pooled odds ratio for all gastrointestinal bleeding was 2.0, with fatal bleeds being very rare. The pooled odds ratio of developing a peptic ulcer on aspirin compared with placebo was 1.3 and that of developing any gastrointestinal symptoms (i.e. nausea, vomiting, heartburn, indigestion) was 1.7.

FACTORS ASSOCIATED WITH INCREASED RISK

Aspirin-related gastrointestinal toxicity is dose related, but no conventionally used prophylactic regimen seems free from risk. Thus a volunteer study of 75 mg aspirin reported treatment-related gastric mucosal bleeding [19] and a recent case control study of patients hospitalized with bleeding peptic ulcer confirmed that risk was increased by all doses of aspirin from 75 mg to 300 mg daily compared with hospital and community controls [20]. The increased risk was not explained by confounding influences of age, sex, prior ulcer history or dyspepsia, or concurrent use of non-aspirin non-steroidal anti-inflammatory drugs, although the latter combination roughly doubled risk. It was estimated that about 900 of the 10 000 episodes of bleeding ulcer which occur in people over 60 years in England and Wales each year could be associated with and ascribed to prophylactic aspirin use. Use of 75 mg rather than 300 mg doses would reduce this increased risk by about 40%.

Patients at greatest risk of developing gastrointestinal toxicity are older patients aged over 60 years, those with a past history of peptic ulceration and those also taking corticosteroids or another non-steroidal anti-inflammatory drug. Cigarette smoking, caffeine and alcohol may also increase susceptibility [21]. The presence of *Helicobacter pylori* does not appear to increase risk, at least in arthritic patients receiving chronic non-steroidal anti-inflammatory therapy [22]. As the background incidence of upper gastrointestinal bleeding rises so steeply with age, even a moderate rise in relative risk in elderly users represents an important increased hazard for them [23].

REDUCING RISK OF GI EFFECTS

As well as using the minimally effective therapeutic dose, administration of aspirin after meals and with adequate fluid may improve tolerance and there is some evidence that enteric-coated and buffered preparations may reduce symptoms, if not limit mucosal damage [24]. Co-administration of H2 receptor antagonists or the prostaglandin analogue misoprostol may offer some degree of prophylaxis against aspirin-induced gastroduodenal damage and should be considered in high-risk patients with strong indications for aspirin therapy. Although specific data concerning aspirin are not available, H2 receptor antagonists can prevent duodenal (but not gastric) lesions associated with non-steroidal anti-inflammatory drug use [25] and omeprazole reduces formation of gastric ulcers [26]. Co-administration of misoprostol with high-dose aspirin can certainly protect against gastric and duodenal lesions (but not dyspepsia) in patients with rheumatoid arthritis [27]. A daily dose as low as 100 mg misoprostol was more effective than placebo in preventing gastric haemorrhagic lesions induced by a four-week course of aspirin 300 mg daily [28].

Intracranial bleeding

The most serious adverse effect associated with aspirin therapy is intracranial haemorrhage. Fortunately this is extremely uncommon and is generally far outweighed by the reduction in the risk of thrombotic episodes, which without aspirin treatment are likely to be far more frequent. The Antiplatelet Trialists' Collaboration reported a non-significant 20% increase in haemorrhagic stroke associated with antiplatelet therapy, with an overall incidence of 0.3% [29]. Unlike the risk of extracranial haemorrhage, primary intracranial haemorrhage does not seem to be significantly influenced by the aspirin dosage used. Neither the Dutch TIA Trial, which compared 30 mg and 283 mg [30], nor the UK TIA Trial, which compared 1200 mg and 300 mg [31], reported different rates of haemorrhage, though the wide confidence intervals do not completely rule out the possibility of a dose relationship. The attributable risk of cerebral haemorrhage (or haemorrhagic transformation) after acute stroke is about 2.0 per 1000 patients.

Hypersensitivity

Estimates of the incidence of aspirin hypersensitivity vary widely, depending on the patient population studied and the definitions used (Table 7.2). Low

Patient population	Incidence of aspirin hypersensitivity (%)
General population	0.3
Rhinitis	1.4
Asthma	4–19
Nasal polyps	14–23
Chronic urticaria	23–28

Table 7.2
Estimated frequency of aspirin hypersensitivity in various patient groups (from reference [32] with permission)

frequencies tend to be associated with studies using historical data, whereas higher frequencies are found after provocative aspirin challenge of patients at risk [32,33]. In a two-year study of 781 asthmatics, drugs were found to precipitate asthma in 10.5% of patients, with aspirin being the agent most commonly held responsible, implicated in nearly half of all drug-induced cases [34].

CLINICAL PRESENTATION

There are two subgroups of aspirin-sensitive subjects. One type develops a respiratory reaction, with rhinitis, asthma or nasal polyps. The other develops a cutaneous reaction, with urticaria, wheals, angioneurotic oedema, hypotension, shock and syncope [32,35,36]. Combined respiratory and cutaneous reactions are uncommon.

The association between aspirin and asthma and nasal polyps was first described by Samter and Beer in 1968 and is sometimes termed Samter's syndrome or triad [37]. Patients usually start with an upper respiratory virus-like syndrome with inflammation of nasal mucosa and paranasal sinuses, developing into a chronic eosinophilic infiltration and eventually leading to nasal polyposis. The response to aspirin typically develops 20 minutes to two hours after ingestion, with symptoms of nasal congestion, facial flushing and mild conjunctivitis. After a few years, asthma develops and occasionally there may be a more severe reaction with headache, vomiting and respiratory failure. The reaction is most commonly seen in middle-aged subjects, without any relevant family history, with normal IgE and eosinophil levels and negative response to skin testing. The underlying mechanism is not understood, but cannot be classified as a true atopic allergic reaction [38].

The cutaneous reaction often develops in people who have taken aspirin previously without problems and avoidance of aspirin does not prevent the reaction recurring. It is therefore thought that in these patients the drug interacts with the underlying urticarial process rather than directly and independently causing it [36]. In a minority of cases, urticaria is exclusively precipitated by aspirin ingestion and it is assumed that these patients are manifesting an IgE-mediated immune response.

MECHANISMS UNDERLYING HYPERSENSITIVITY

The pathogenesis of most cases of aspirin hypersensitivity is believed to relate to imbalance between the rates of enzymatic synthesis of prostaglandins (which are bronchodilators and may suppress T cell-mediated inflammatory responses in the lung) and leukotrienes (which are bronchoconstrictors) from arachidonic acid. Inhibition of COX activity will divert arachidonic acid to the 5-lipoxygenase pathway and result in an increase in leukotriene generation. Aspirin-sensitive subjects produce exaggerated amounts of leukotrienes from mast cells into the nasal secretions in response to aspirin challenge [39,40], but levels in bronchoalveolar lavage fluid were not greatly raised [41]. However, specific leukotriene receptor antagonists are highly effective in preventing aspirin-induced asthma [42], as is the 5–lipoxygenase inhibitor Zileuton [43]. Complement activation [44] and abnormal platelet response [45] have also been suggested as possible pathophysiological mechanisms.

Patients who are sensitive to aspirin may develop cross-sensitivity to other non-steroidal anti-inflammatory drugs and non-narcotic analgesics. The degree of cross-sensitivity seems to be related to the degree to which agents inhibit COX, rather than their structural relationship [46]. Thus non-acetylated salicylate salts seem to be much less likely to produce hypersensitivity reactions. Cross-reaction may even extend to natural salicylates present in foods, such as bananas, apples, rhubarb, dried fruits and peas. However, the cross-reaction is no longer believed to extend to azo dyes, such as tartrazine [41].

MANAGEMENT

Confirmation of aspirin hypersensitivity requires placebo-controlled challenge tests, using oral or inhaled aspirin preparations [46–48]. This should only be carried out in experienced centres with appropriate facilities for monitoring and treatment. Management is non-specific, relying mainly on beta-adrenergic agonists to treat bronchospasm and topical vasoconstrictors for naso-ocular symptoms and signs. Antihistamines and mast cell stabilizers, such as ketotifen, may be useful, but response to corticosteroids may be disappointing. Antibiotics may be required to treat infective exacerbations of sinusitis. The leukotriene receptor antagonists and 5-lipoxygenase inhibitors are likely to play an important role in the future. Surgical intervention for chronic sinusitis and nasal polyps may be useful [36,48].

Avoidance of aspirin preparations (and of other analgesic and non-steroidal anti-inflammatory drugs which cross-react with aspirin) is obviously of central importance. Patients with proven sensitivity must be warned to avoid all aspirin-like preparations, including over-the-counter remedies, and in severe cases a 'salicylate-free' diet may be necessary. Aspirin desensitization may be useful in protecting patients with aspirin-induced respiratory and nasal symptoms, but can precipitate severe bronchospasm and only remains effective while the patient is regularly taking aspirin. It is ineffective in patients with the cutaneous urticarial syndrome [49].

Reye's syndrome

Reye's syndrome is a rare childhood illness presenting as acute encephalopathy and hepatic dysfunction which develops after an apparently trivial viral infection [50]. It commonly leads to death or severe permanent neurological damage. The syndrome is believed to arise consequent to a virus–host interaction, occurring in a genetically susceptible child and possibly modified by an exogenous agent, notably aspirin. Many apparent cases are now recognized to be associated with inborn errors of metabolism [51,52]. Liver biopsy is diagnostic and shows severe microvesicular panlobular fatty infiltration, with swollen and irregular mitochondria visible on electron microscopy [53]. The mitochondrial dysfunction appears to be of critical importance, with normal beta oxidation blocked at multiple sites, leading to accumulation of acyl CoA esters, reduced energy and ATP production and prevention of proper enzyme processing on the mitochondrial inner membrane [54]. It may be relevant that in animal experiments salicylic acid has been shown to inhibit beta oxidation of long chain fatty acids and mitochondrial activation [55], although in human volunteer studies aspirin has not been shown to alter fatty acid oxidation [56].

EVIDENCE TO IMPLICATE ASPIRIN

Aspirin consumption was first implicated as a possible causative factor in a series of case control studies conducted in America in the early 1980s [57–59]. Although criticized for their poor design, with lack of diagnostic confirmation, poor control for severity, type and viral cause of the prodrome and bias in recall and data collection [60,61], subsequent studies appeared to confirm the association. Thus the important US Public Health Laboratory Service studies showed a 16- to 26-fold increase in relative risk of developing the syndrome if aspirin had been taken [62,63]. A later well-designed, multicentre study reported a dose-related effect, with up to a 35-fold increase in relative risk [64]. In Britain, epidemiological evidence also seemed to implicate antipyretic use, and especially aspirin rather than paracetamol, in the development of Reye's syndrome, with an especially high incidence in Northern Ireland [65].

In the light of the growing evidence indicating that aspirin might be a contributory factor in causing some cases of Reye's syndrome, the British Committee on Safety of Medicines (CSM) recommended in 1986 that aspirin should not be given to children aged under 12 years of age, unless specifically indicated for childhood rheumatic conditions [66]. This was reinforced by a public information campaign and labelling changes and the withdrawal of all paediatric preparations of aspirin. Careful surveillance of the numbers of cases of Reye's syndrome before and after the CSM advice showed a marked fall [67–69], as had previously been observed with decreased aspirin use in studies in America [70] (Figure 7.1).

Not all authors accept that there is a causal link between aspirin and Reye's syndrome, highlighting the apparent decline in Reye's before the reduction in aspirin use [71] and case control studies conducted in Japan [72] and Australia [73] have failed to confirm any relationship. Nevertheless, given the accumulated weight of the evidence, together with the serious nature of the condition and the availability of alternative effective remedies (paracetamol), a recent editorial concluded that it was vital that the dangers of using aspirin in children with common viral illnesses must be kept before the public [52]. Countries where aspirin is still given to children are being encouraged to withdraw paediatric preparations from sale [74].

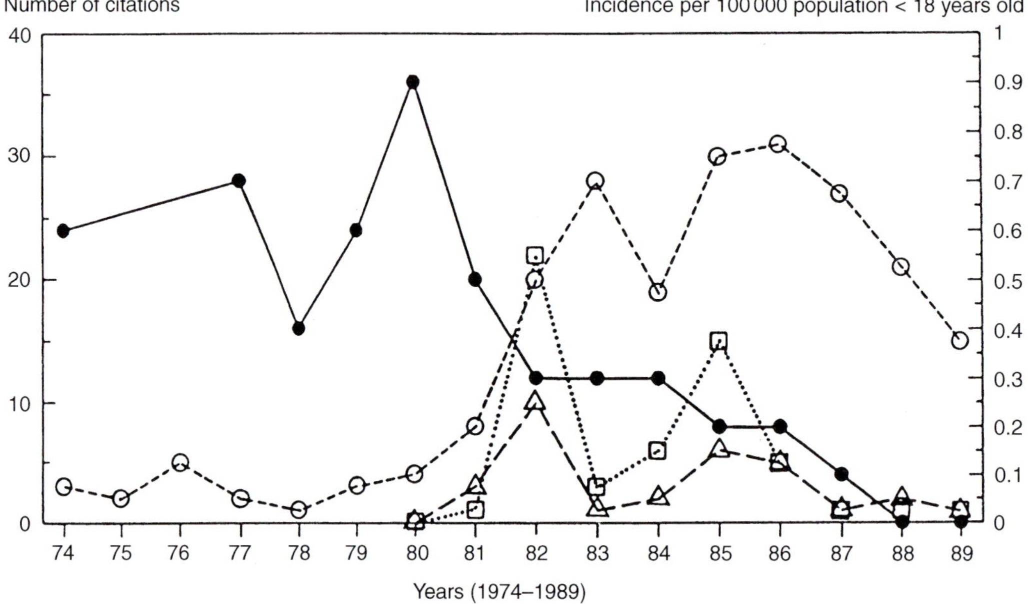

Fig. 7.1
Trend in number of reported cases of Reye's syndrome (solid line) and of citations in the medical literature (○) and in newspapers (□) and magazines (△) in the USA between 1974 and 1989 (from reference [68] with permission).

Other adverse effects

NEUTROPENIA, THROMBOCYTOPENIA AND HAEMOLYSIS

There have been occasional reports of aspirin-induced thrombocytopenia and very rarely of agranulocytosis and aplastic anaemia [75]. It may also lead to haemolysis in patients with glucose-6-phosphate dehydrogenase deficiency [76].

EFFECTS ON LIVER FUNCTION

Mild elevation of transaminase enzymes and, less commonly, of alkaline phosphatase occasionally develops after 1–4 weeks of aspirin therapy, appears to be dose related, is usually transient or reversible on stopping the drug and is most commonly seen in patients with active juvenile rheumatoid arthritis or systemic lupus erythematosus, elderly patients or those who have impaired renal function [77]. It does not seem to occur if serum salicylate concentrations are less than 200 mg/l [78]. In a comparative study in arthritic patients, aspirin was significantly more hepatotoxic than ibuprofen [79].

Symptomatic hepatitis and jaundice are rare, but some patients may develop nausea, vomiting, anorexia, abdominal pain, liver tenderness and/or hepatomegaly. Severe liver injury has occasionally been reported [80]. The histopathological appearances are of areas of scattered necrosis, mononuclear cell infiltration of the portal tracts and biliary stasis. It is recommended that aspirin treatment should not be used or should be stopped if hepatic enzyme levels exceed three times the upper limit of normal or if there is other evidence of hepatic damage [81].

EFFECTS ON RENAL FUNCTION

There is conflicting evidence on the ability of aspirin as a single agent in therapeutic doses to cause acute or chronic renal failure [82,83]. In healthy adults, aspirin probably has no significant adverse effects on renal function. In predisposed individuals with established chronic renal insufficiency or certain connective tissue disorders, short-term therapeutic use of aspirin may precipitate reversible acute renal failure. Only one case control study has demonstrated a low but statistically significant risk of endstage renal disease in association with aspirin therapy, after correction for other analgesics, but the chronic use of combinations of aspirin with other analgesics may lead to papillary necrosis and interstitial nephritis ('analgesic nephropathy') [84]. This was a major problem in many Western countries until the 1970s, when greater awareness of the problem led to sales restrictions and a dramatic fall in the availability of analgesic mixtures, especially those containing phenacetin [85].

Toxicity

Aspirin intoxication may arise intentionally, following deliberate suicidal or parasuicidal overdose, or unintentionally following therapeutic use of too high a dose or in young children who have accidentally swallowed tablets. Until the 1970s, salicylates were the most common cause of self-poisoning, with over 10 000 hospital admissions in England and Wales each year and 300 deaths [86]. Changes in prescribing habits, particularly use of lower therapeutic doses for antithrombotic prophylaxis, avoidance in children because of concerns about Reye's syndrome and the increased use of alternative household analgesics such as paracetamol, together with the introduction of safety packaging and restrictions in pack sizes have all helped to reduce the size of the problem. Nevertheless, in Britain poisoning from aspirin still accounts for around 5000 hospital admissions and 50 deaths each year [87] and there is little evidence from recent studies that the rates of self-poisoning with aspirin are continuing to fall [88].

Relatively small overdoses of aspirin can cause significant problems, especially in the very young and in the old who are more likely to have a complicating prior metabolic, renal or pulmonary abnormality. The threshold for more vigorous therapy is therefore at lower salicylate concentrations in children and elderly patients than in healthy adults. The saturable nature of aspirin pharmacokinetics means that accumulation can rapidly develop. In adults, the fatal dose is generally estimated at about 500 mg/kg, but there are reports of survival after much larger amounts, even up to 130 g [89]. In children, doses as little as 4 g in a four-year-old have proved fatal. Salicylate toxicity may be enhanced if paracetamol or ibuprofen is taken concurrently. Mortality appears to be much higher in chronic salicylate intoxication than after acute overdose [90].

SYMPTOMS OF TOXICITY

Mild chronic salicylate intoxication is termed salicylism and may start to develop with plasma salicylate levels above 200 mg/l (1.44 mmol/l). Symptoms include nausea and vomiting, epigastric discomfort, headaches, tinnitus and dizziness. Such symptoms were long used as a guide to achieve adequate anti-inflammatory dosage [91]. More severe intoxication may lead to flushing and sweating, hyperventilation, thirst and a variety of central nervous system disturbances including high tone deafness, impaired vision, tremor, delirium and drowsiness. Severe salicylate poisoning is characterized by gross disturbance of acid–base, electrolyte and fluid balance [92,93]. Stimulation of the respiratory centre causes hyperventilation and, in adults, a respiratory alkalosis and there is compensatory excretion of bicarbonate, water and electrolytes in urine, leading to a metabolic acidosis, potentially severe dehydration, hypokalaemia and hypo- or hypernatraemia. In young children, respiratory alkalosis is uncommon but metabolic acidosis may rapidly develop. Central nervous system effects, such as marked confusion, reduced conscious level leading to coma and convulsions, are associated with acidosis, which enhances transfer of salicylate across the blood–brain barrier. Hyperpyrexia is a common complication in children. Uncoupling of oxidative phosphorylation reduces ATP production and increases oxygen utilization and carbon dioxide and lactate production. Hypo- or, less commonly, hyperglycaemia may develop, as may hypoprothrombinaemia with consequent bleeding complications, although haematemesis is rare. Renal failure may intervene, preventing the excretion of both the products of elimination and the drug and its metabolites and pulmonary oedema and cardiovascular collapse are poor prognostic signs. Death usually occurs secondary to cardiorespiratory arrest [92].

At post mortem the stomach will often still contain residual aspirin and the gastric mucosa may be eroded and haemorrhagic. Petechial haemorrhages are prominent on the skin and parietal pleura and pulmonary oedema may be present, despite absence of evidence of cardiac disease. Salicylate will be detectable in blood and urine.

MANAGEMENT

Treatment of aspirin toxicity should be directed at removing salicylate from the body and minimizing its effects [94] (Figure 7.2). Aspirin treatment should be stopped as soon as toxicity is suspected. Current recommendations for managing aspirin poisoning and advice on treatment of specific patients can be obtained day and night in Britain from Poisons Information Centres. Telephone numbers are printed in the British National Formulary, MIMS and in local telephone directories. The following is based on information supplied by them.

Except in exceptional circumstances, nearly all patients, including those who appear well, should be admitted to hospital, where laboratory monitoring can take place. Salicylate will be detectable in blood, urine and gastric aspirate. Blood should be taken at four hours postingestion for plasma salicylate estimation and then repeated three-hourly until the peak level has been passed. Early blood concentrations can sometimes be misleadingly low and peak salicylate levels may be delayed for up to 24 hours [95]. Done devised a nomogram which grades the severity of salicylate intoxication by

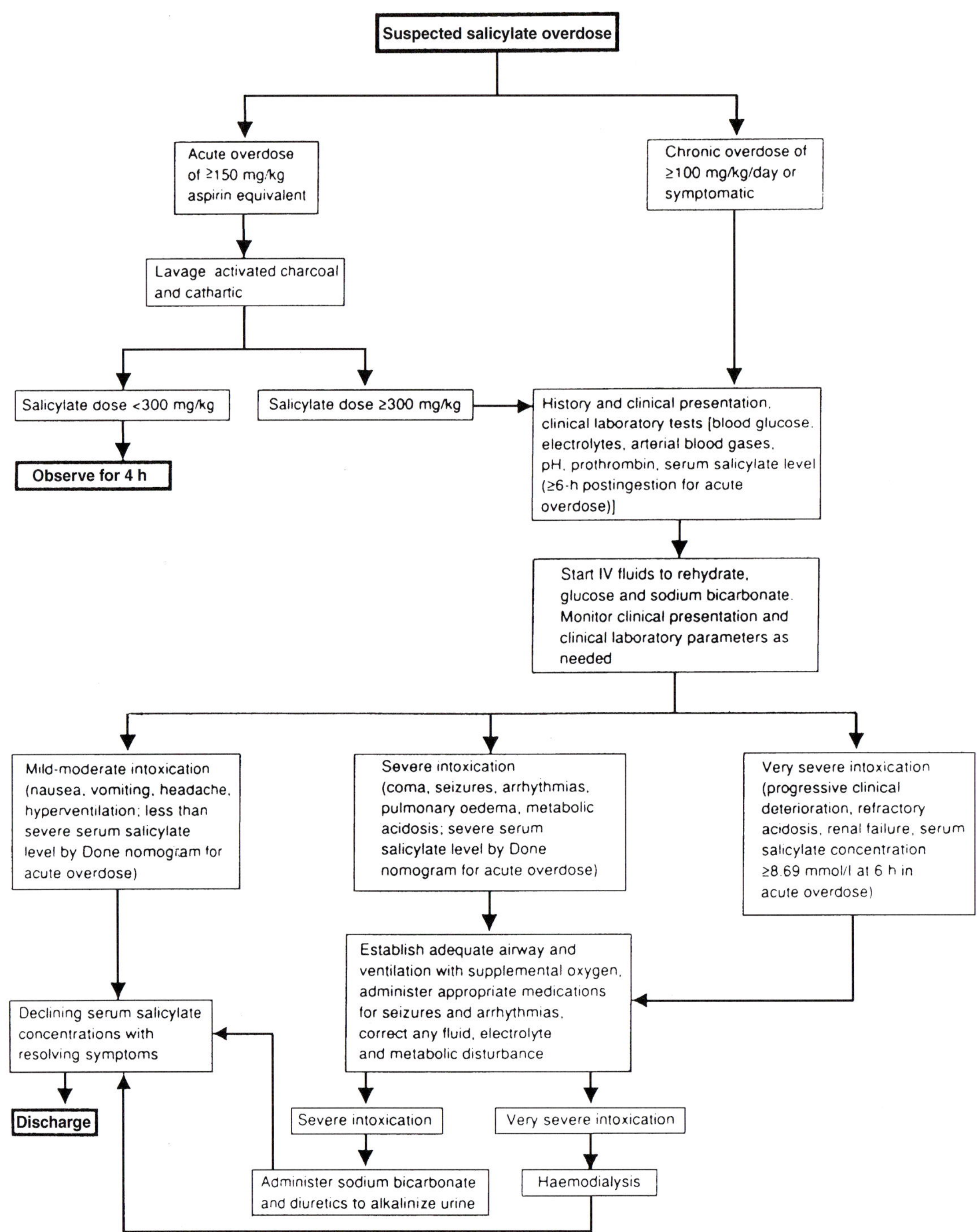

Fig. 7.2
Management of salicylate poisoning in adults (from reference [94] with permission).

relating serum salicylate concentrations to time of aspirin ingestion [96] (Figure 7.3). However, interpretation of salicylate levels is difficult, as they will be influenced not only by time since ingestion but also by aspirin formulation, co-ingestants, metabolic state and age. Regular measurement of urea and electrolytes, plasma glucose, prothrombin time and arterial pH and blood gases should also be made. Optimal medical care requires a close interaction between clinical toxicology laboratory and clinical staff [97].

Because absorption of aspirin from the gastrointestinal tract may be slow, especially following ingestion of enteric-coated preparations [98], measures should always be taken to remove the drug from the gut by induction of emesis or gastric lavage. A worthwhile recovery of salicylates can be achieved up to 24 hours after ingestion. Activated charcoal in a dose of 50–100 g for adults and 1 g/kg in children should be administered up to four hours postingestion for aspirin overdoses more than 120 mg/kg. As desorption may occur, this should be repeated four-hourly until there is a clinical improvement in the patient's condition. Salicylate absorption may be further reduced by administration of a single dose of a rapidly acting cathartic such as sorbitol [99]. This should be given with the first dose of activated charcoal, in a dose of 1 ml/kg of a 70% solution.

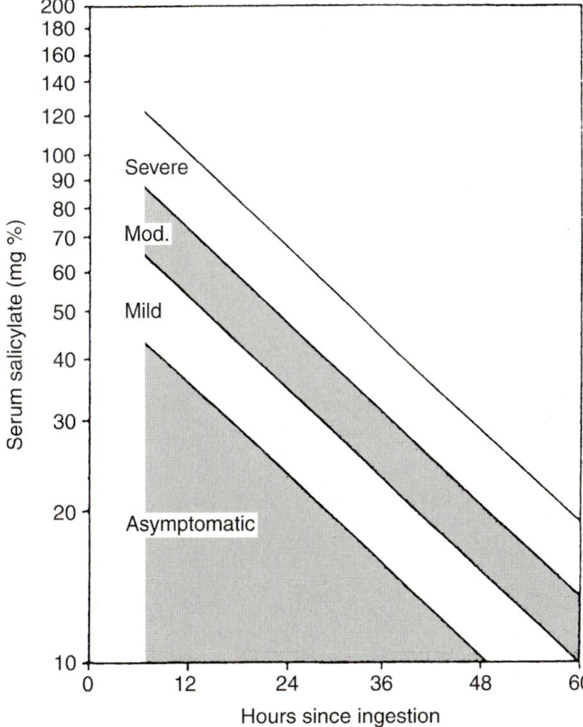

Fig. 7.3
Done nomogram for salicylate intoxication (from reference [96] with permission).

If peak salicylate levels are known to be less than 300 mg/l (2.16 mmol/l) in adults or 250 mg/l (1.8 mmol/l) in a child, with only mild clinical effects, then further treatment is unlikely to be needed. If the salicylate level is greater than 600 mg/l (4.32 mmol/l) in an adult or 450 mg/l (3.24 mmol/l) in a child or old person, or clinical effects are moderate, then urinary alkalization to raise the urine pH above 7.5 should be instituted. In all cases, dehydration should be promptly corrected with oral or intravenous fluids, guided by the results of laboratory tests and in more severe cases by monitoring of central venous pressure. Large volumes of fluid may be required. Acidosis, which results in a shift of plasma salicylate into brain or other tissues, should be urgently corrected by infusion of sodium bicarbonate in sufficient quantity to maintain an alkaline diuresis, which greatly enhances urinary excretion of salicylate.

Correction of hypoglycaemia and ketosis will require glucose administration and hypokalaemia (which prevents urinary excretion of alkali) should be corrected by supplementing fluids with potassium chloride once adequate renal function has been confirmed. A typical regimen would be 1 litre of 1.26% (isotonic) sodium bicarbonate plus 40 mmol of potassium chloride infused over four hours in adults and 1 ml/kg of 8.4% sodium bicarbonate (= 1 mmol/kg) plus 20 mmol of potassium chloride diluted in 0.5 litres of dextrose saline infused at 2–3 ml/kg/h in children. Prolongation of prothrombin time may be corrected by administration of vitamin K. Great care is needed with the use of centrally depressant drugs, as asystole has been reported in patients with salicylate intoxication given intravenous diazepam [100].

Forced alkaline diuresis was for long the approved treatment for salicylate overdose, but is no longer recommended. It has not been shown to be superior to use of alkali alone and is more likely to cause serious electrolyte disturbance and fluid overload [101].

In severe intoxication, haemodialysis may become necessary. It should be considered in patients with very high plasma salicylate concentrations and in those with acid–base or electrolyte disturbance resistant to correction, in those with established cardiorespiratory or renal failure, in those with convulsions and central nervous system effects not responsive to correction of acidosis and in those whose condition continues to deteriorate despite appropriate therapeutic measures.

Conclusions

It is often said that aspirin would not be registered by the licensing authorities if it were submitted as a new drug today, because of its adverse effect profile. In fact, considering the widespread use of aspirin, its safety record in clinical use is remarkably good. Concern about its gastrointestinal toxicity must be kept in proportion. For example, the attributable risk for haematemesis is 0.2–1.0 per 1000 person-years of exposure, whereas in patients with a history of myocardial infarction the absolute reduction in vascular events is 40 per 1000 patients treated [102]. Nevertheless, aspirin overdosage remains a significant problem and its ready availability and proven clinical benefit must not permit complacency to develop over its serious toxicity when misused.

KEYPOINTS

- Side effects from thromboprophylactic doses of aspirin or occasional analgesic or anti-inflammatory doses are generally mild and infrequent.
- Gastrointestinal symptoms are reported most often and are dose related. Bleeding complications may be serious and may develop asymptomatically. Risk is greatest in the old, those with previous peptic ulceration and those taking other anti-inflammatory drugs and alcohol. Intracranial haemorrhage is rare, but often fatal.
- Hypersensitivity reactions are uncommon and manifest as either a respiratory reaction (rhinitis, asthma and nasal polyps) or a cutaneous reaction (urticaria, angioneurotic oedema and shock). There is often cross-reactivity with other NSAIDs.
- An association between aspirin and Reye's syndrome means that aspirin should not be given to children under the age of 12 years, unless specifically indicated.
- Accidental or intentional overdose of aspirin may lead to serious disturbance of acid–base, electrolyte and fluid balance, with cardiorespiratory and renal failure and central nervous system toxicity. Management centres on correction of biochemical abnormalities and measures to reduce absorption and promote excretion of salicylate.

REFERENCES

1. Miller RR, Jick H. Acute toxicity of aspirin in hospitalised medical patients. Am J Med Sci 1977; 274: 271–279.
2. Steering Committee of the Physicians' Health Study Research Group. Final report on the aspirin component of the ongoing Physicians' Health Study. N Engl J Med 1989; 321: 129–135.
3. Alvarez AS, Summerskill WHJ. Gastrointestinal haemorrhage and salicylates. Lancet 1958; ii: 179–182.

4. Coggon D, Langman MJS, Spiegelhalter D. Aspirin, paracetamol and haematemesis and melaena. Gut 1982; 23: 340–344.
5. Graham DY, Lacey Smith J. Aspirin and the stomach. Ann Intern Med 1986; 104: 390–398.
6. Faulkner G, Prichard P, Somerville K, Langman MJS. Aspirin and bleeding peptic ulcers in the elderly. Br Med J 1988; 297: 1311–1313.
7. Skander MP, Ryan FP. Non-steroidal antiinflammatory drugs and pain-free peptic ulceration in the elderly. Br Med J 1988; 297: 833.
8. Leonards JR, Levy G. Reduction or prevention of aspirin induced occult gastrointestinal blood loss in man. Clin Pharmacol Ther 1969; 10: 571–576.
9. Greenberg PD, Cello JP, Rockey DC. Asymptomatic chronic gastrointestinal blood loss in patients taking aspirin or warfarin for cardiovascular disease. Am J Med 1996; 100: 598–604.
10. Douthwaite AH, Lintott GAM. Gastroscopic observation of the effect of aspirin and certain other substances on the stomach. Lancet 1938; ii: 1222–1225.
11. Caruso I, Bianchi Porro G. Gastroscopic evaluation of anti-inflammatory agents. Br Med J 1980; 1: 75–78.
12. Hawkey CJ. Nonsteroidal antiinflammatory drugs and peptic ulcers. Br Med J 1990; 300: 278.
13. Lanas A, Hirschowitz BI. Significant role of aspirin in patients with oesophagitis. J Clin Gastroenterol 1991; 13: 622–627.
14. Morgan G. Minimising the risks of NSAID pill oesophagitis. Int J Gastroenterol 1997; 1: 14–16.
15. Editorial. Aspirin and the stomach. Br Med J 1981; 282: 91–92.
16. Levy M. Aspirin use in patients with major upper gastrointestinal bleeding and peptic ulcer disease. N Engl J Med 1974; 290: 1158–1162.
17. Henry D, Lim LLY, Garcia-Rodriguez LA *et al.* Variability in risk of gastrointestinal complications with individual nonsteroidal antiinflammatory drugs: results of a collaborative meta-analysis. Br Med J 1996; 312: 563–566.
18. Roderick PJ, Wilkes HC, Meade TW. The gastrointestinal toxicity of aspirin: an overview of randomised trials. Br J Clin Pharmacol 1993; 35: 219–226.
19. Pritchard PJ, Kitchingman GK, Walt RP, Daneshmend TK, Hawkey CJ. Human gastric mucosal bleeding induced by low dose aspirin but not warfarin. Br Med J 1989; 298: 493–496.
20. Weil J, Colin-Jones D, Langman M *et al.* Prophylactic aspirin and risk of peptic ulcer bleeding. Br Med J 1995; 310: 827–830.
21. Polisson R. Nonsteroidal anti-inflammatory drugs: practical and theoretical considerations in their selection. Am J Med 1996; 100 (suppl 2A): 31S–36S.
22. Kim JG, Graham DY, Misoprostol Study Group. Helicobacter pylori infection and development of gastric or duodenal ulcer in arthritic patients receiving chronic NSAID therapy. Am J Gastroenterol 1994; 89: 203–207.

23. Bateman DN, Kennedy JG. Non-steroidal anti-inflammatory drugs and elderly patients. Br Med J 1995; 310: 817–818.
24. Anonymous. Which prophylactic aspirin? Drugs Therapeut Bull 1997; 35: 7–8.
25. Langman MJS. Treating ulcers in patients receiving anti-arthritic drugs. Q J Med 1989; 73: 1089–1091.
26. Walan A, Bader JP, Classen M *et al.* Effect of omeprazole and ranitidine on ulcer healing and relapse rates in patients with benign gastric ulcer. N Engl J Med 1989; 320: 69–75.
27. Roth S, Agarwal N, Mahowald M *et al.* Misoprostol heals gastroduodenal injury in patients with rheumatoid arthritis receiving aspirin. Arch Intern Med 1989; 149: 775–779.
28. Goddard AF, Donnelly MT, Filipowicz B *et al.* Low dose misoprostol as prophylaxis against low-dose aspirin-induced gastroduodenal mucosal injury. Gut 1996; 39: A33.
29. Antiplatelet Trialists' Collaboration. Collaborative overview of randomised controlled trials of antiplatelet therapy – I: prevention of death, myocardial infarction, and stroke by prolonged antiplatelet therapy in various categories of patients. Br Med J 1994; 308: 81–106.
30. Dutch TIA Trial Study Group. A comparison of two doses of aspirin (30mg vs. 283mg a day) in patients after a transient ischaemic attack or minor ischaemic stroke. N Engl J Med 1991; 325: 1261–1266.
31. Farrell B, Godwin J, Richards S, Warlow C. The United Kingdom transient ischaemic attack (UK-TIA) aspirin trial: final results. J Neurol Neurosurg Psychiatr 1991; 54: 1044–1054.
32. Settipane GA. Aspirin and allergic diseases. Am J Med 1983; 74: 102–109.
33. Clissold SP. Aspirin and related derivatives of salicylic acid. Drugs 1986; 32 (suppl 4): 8–26.
34. Iamandescu IB. NSAID-induced asthma: peculiarities related to background and association with other drug or non-drug etiological agents. Allergol Immunopathol 1989; 17: 285–290.
35. Szczeklik A. Antipyretic analgesics and the allergic patient. Am J Med 1983; 75: 82–84.
36. Stevenson DD, Simon RA. Aspirin sensitivity: respiratory and cutaneous manifestations. In: Middleton E Jr, Reed CE, Ellis EF (eds) Allergy: Principles and Practice. St Louis: CV Mosby, 1993: 1747–1767.
37. Samter M, Beer RF. Intolerance to aspirin: clinical studies and consideration of its pathogenesis. Ann Intern Med 1968; 68: 975–983.
38. Stevenson DD, Lewis RA. Proposed mechanisms of aspirin sensitivity reactions. J Allergy Clin Immunol 1987; 80: 788–790.
39. Kowalski ML, Sliwinska-Kowalska M, Igarishi Y *et al.* Nasal secretions in response to acetylsalicylic acid. J Allergy Clin Immunol 1993; 91: 580–598.
40. Fischer AR, Rosenberg MA, Lilly CM *et al.* Direct evidence for a role of the mast cell in the response to aspirin in the aspirin-sensitive patient. J Allergy Clin Immunol 1994; 94: 1046–1056.

41. Sladek K, Dworski R, Soja J *et al.* Eicosanoids in bronchoalveolar lavage fluid of aspirin-intolerant patients with asthma after aspirin challenge. Am J Respir Crit Care Med 1994; 149: 940–946.
42. Dahlen B, Kumlin M, Margolskee DJ *et al.* The leukotriene receptor antagonist MK-0679 blocks airway obstruction induced by bronchial provocation with lysine aspirin in aspirin-sensitive asthmatics. Eur Respir J 1993; 6: 1018–1026.
43. Israel E, Fischer AR, Rosenberg MA *et al.* The pivotal role of 5-lipoxygenase products in the reaction of aspirin sensitive asthmatics to aspirin. Am Rev Respir Dis 1993; 148: 1447–1451.
44. Pleskow WW, Chenoweth DE, Smith RA, Stevenson DD, Curd JG. The absence of detectable complement activation in aspirin sensitive patients during aspirin challenge. J Allergy Clin Immunol 1983; 72: 462–468.
45. Ameisen JC, Capron A, Joseph M *et al.* Aspirin-sensitive asthma: abnormal platelet response to drugs inducing asthma attacks. Diagnostic and pathophysiological implications. Int Arch Allergy Appl Immunol 1985; 78: 438–448.
46. Szczelik A, Gryglewski RJ, Czerniawska-Mysik G. Relationship of inhibition of prostaglandin biosynthesis by analgesics to asthma attacks in aspirin-sensitive patients. Br Med J 1975; 1: 67–69.
47. Fischer AR, Israel E. Identifying and treating aspirin-induced asthma. J Respir Dis 1995; 16: 304–317.
48. Geba GP. Aspirin and exercise-induced asthma. In: Fishman AP, Elias JA, Fishman JA *et al* (eds) Fishman's Pulmonary Diseases and Disorders, 3rd edn. New York: McGraw Hill, 1998: 745–755.
49. Pleskow WW, Stevenson DD, Mathison DA *et al.* Aspirin desensitisation in aspirin-sensitive asthmatic patients: clinical manifestations and characterization of the refractory period. J Allergy Clin Immunol 1982; 69: 11–19.
50. Reye RDK, Morgan G, Baral J. Encephalopathy and fatty degeneration of the viscera: a disease entity in childhood. Lancet 1963; ii: 749–752.
51. Gauthier M, Guay J, Lacroix J, Lortie A. Reye's syndrome: a reappraisal of diagnosis in 49 presumptive cases. Am J Dis Child 1989; 143: 1181–1185.
52. Glasgow JFT, Moore R. Reye's syndrome 30 years on: possible marker of inherited metabolic disorders. Br Med J 1993; 307: 950–951.
53. Treem WR, Witzleben CA, Piccoli DA *et al.* Medium chain and long chain acyl CoA dehydrogenase deficiency: clinical, pathological and ultrastructural differentiation from Reye's syndrome. Hepatology 1986; 6: 1270–1278.
54. Van Cosker RN, De Vivo DC, Blake D *et al.* Adult Reye's syndrome: a review of new evidence for a generalised defect in intramitochondrial enzyme processing. Neurology 1991; 41: 1815–1821.
55. Deschamps D, Fisch C, Fromenty B *et al.* Inhibition by salicylic acid of the activation and thus oxidation of long chain fatty acids. Possible role in the development of Reye's syndrome. J Pharmacol Exp Ther 1991; 259: 894–904.

56. Williams FM, Ferner RE, Graham M *et al.* The metabolic effects of aspirin in fasting and fed subjects: relevance to the aetiology of Reye's syndrome. Eur J Clin Pharmacol 1990; 38: 519–521.
57. Starko KM, Ray CG, Dominguez LB, Stromberg WL, Woodall DF. Reye's syndrome and salicylate use. Pediatrics 1980; 66: 859–864.
58. Halpin TJ, Holtzhauer FJ, Campbell RJ *et al.* Reye's syndrome and medication use. JAMA 1982; 248: 687–691.
59. Waldman RJ, Hall WN, McGee H, Van Amburg G. Aspirin as a risk factor in Reye's syndrome. JAMA 1982; 247: 3089–3094.
60. Daniels SR, Greenberg RS, Ibrahim MA. Scientific uncertainties in the studies of salicylate use and Reye's syndrome. JAMA 1983; 249: 1311–1316.
61. Hall SM. Reye's syndrome and aspirin: a review. J Roy Soc Med 1986; 79: 596–598.
62. Hurwitz ES, Barrett MJ, Bregman D *et al.* Public Health Service Study on Reye's syndrome and medication: report of the pilot phase. N Engl J Med 1985; 313: 849–857.
63. Hurwitz ES, Barrett MJ, Bregman D *et al.* Public Health Service Study on Reye's syndrome and medication: report of the main study. N Engl J Med 1987; 257: 1905–1911.
64. Forsyth BW, Hurwitz RA, Acampora D *et al.* New epidemiological evidence confirming that bias does not explain the aspirin/Reye's syndrome association. JAMA 1989; 261: 2517–2524.
65. Hall SM, Plaster PA, Glasgow JFT, Hancock P. Preadmission antipyretics in Reye's syndrome. Arch Dis Child 1988; 63: 857–866.
66. Anonymous. CSM Update: Reye's syndrome and aspirin. Br Med J 1986; 292: 1590.
67. Newton L, Hall SM. Reye's syndrome in the British Isles: report for 1990–91 and the first decade of surveillance. Communicable Dis Rep 1993; 3: R11–16.
68. Soumeral SB, Ross-Degnan D, Kahn JS. Effects of professional and media warnings about the association between aspirin use and Reye's syndrome. Milbank Quart 1992; 70: 155–182.
69. Casteels-Van Daek M, Van Geet C, Wouters K, Eggermont E. The changing clinical pattern of Reye's syndrome 1982–1990. Arch Dis Child 1996; 74: 400–405.
70. Barrett MJ, Hurwitz ES, Shonbergen LB, Rogers MF. Changing epidemiology of Reye syndrome in the United States. Pediatrics 1986; 77: 598–602.
71. Shindell S. Reduction of deaths after drug labelling for risk of Reye's syndrome. Lancet 1993; 341: 119.
72. Committee on Reye's Syndrome Research. Japanese Ministry of Health and Welfare. Official Report, 1983.
73. Orlowski JP, Campbell P, Goldstein S. Reye's syndrome: a case control study of medication use and associated viruses in Australia. Cleveland Clinic J Med 1990; 57: 323–329.
74. Larsen SV. Reye's syndrome. Med Sci Law 1997; 37: 235–241.

75. Cuthbert MF. Adverse reactions to non-steroidal antirheumatic drugs. Curr Med Res Opin 1974; 2: 600–609.
76. Magee P, Beeley L. Drug-induced blood dyscrasias. Pharm J 1991; 246: 396–397.
77. Day RO. Aspirin and salicylates. In: Kelley WN, Harris ED, Ruddy S (eds) Textbook of Rheumatology, 4th edn. New York: WB Saunders, 1994: 681–691.
78. Zimmerman HJ. Effects of aspirin and acetaminophen on the liver. Arch Intern Med 1981; 141: 333–342.
79. Freeland GR, Northington RS, Hedrich DA, Walker BR. Hepatic safety of two analgesics used over the counter: ibuprofen and aspirin. Clin Pharmacol Ther 1988; 43: 473–479.
80. Lewis JH. Hepatic toxicity of nonsteroidal antiinflammatory drugs. Clin Pharmacol 1984; 3: 128–138.
81. Paulus HE. FDA Advisory Committee Meeting. Arthritis Rheum 1983; 6: 206–214.
82. Perneger TV, Whelton PK, Klag MJ. Risk of kidney failure associated with the use of acetaminophen, aspirin and nonsteroidal anti-inflammatory drugs. N Engl J Med 1994; 331: 1675–1679.
83. D'Agati V. Does aspirin cause acute or chronic renal failure in experimental animals and in humans? Am J Kidney Dis 1996; 28 (suppl 1): S24–S29.
84. Prescott LF. Analgesic nephropathy. Drugs 1982; 23: 75–149.
85. Rainsford KD. Aspirin and the Salicylates. London: Butterworths, 1984.
86. Graham JDP. Acute poisoning. In: Bodley Scott R (ed.) Price's Textbook of the Practice of Medicine, 12th edn. Oxford: Oxford University Press, 1978.
87. Committee on Safety of Medicines/Medicines Control Agency. Paracetamol and aspirin. Curr Prob Pharmacovigilance 1997; 23: 9.
88. McLoone P, Crombie IK. Hospitalisation for deliberate self-poisoning in Scotland from 1981 to 1993: trends in rates and types of drugs used. Br J Psychiatr 1996; 169: 81–85.
89. Insel PA. Analgesic–antipyretic and antiinflammatory agents: the salicylates. In: Hardman JG, Limbird LE (eds) Goodman and Gilman's The Pharmacological Basis of Therapeutics, 9th edn. New York: Mosby, 1996: 617–657.
90. Proudfoot AT. Toxicity of salicylates. Am J Med 1983; 75: 99–103.
91. Graham JDP, Parker WA. The toxic manifestations of sodium salicylate therapy. Q J Med 1948; 17: 153–163.
92. Jewett JF. Salicylate poisoning. N Engl J Med 1973; 288: 967–968.
93. Henry J. ABC of poisoning. Analgesic poisoning: I. Salicylates. Br Med J 1984; 289: 820–822.
94. Klein-Schwartz W, Oderda GM. Poisoning in the elderly. Drugs Aging 1991; 1: 67–89.
95. Anonymous. Salicylate poisoning. Lancet 1981; 2: 130.
96. Done AK. Salicylate intoxication: significance of measurements of salicylate in blood in cases of acute ingestion. Pediatrics 1960; 26: 800–807.

97. Krause DS, Wolf BA, Shaw LM. Acute aspirin overdose: mechanisms of toxicity. Ther Drug Monit 1992; 14: 441–451.
98. Pierce RP, Gazewood J, Blake RL Jr. Salicylate poisoning from enteric-coated aspirin. Delayed absorption may complicate management. Postgrad Med 1991; 89: 61–62.
99. Keller RE, Schwab RA, Krenzelok EP. Contribution of sorbitol combined with activated charcoal in prevention of salicylate absorption. Ann Emerg Med 1990; 19: 654–656.
100. Berk WA, Anderson JC. Salicylate-associated asystole: report of two cases. Am J Med 1989; 86: 505–506.
101. Prescott LF, Balali-Mood M, Critchley JA, Johnstone AF, Proudfoot AT. Diuresis or urinary alkalinisation for salicylate poisoning. Br Med J 1982; 285: 1383–1386.
102. Symmons DPM. Safety profile of low-dose aspirin. Lancet 1996; 348: 1394–1395.

Contraindications and drug interactions

Introduction

The great majority of people can take therapeutic doses of aspirin safely and without adverse effect. However, there are some definite contraindications to its use, even in low dosage, and special thought and care need to be given to its use in certain clinical situations and with particular patient groups [1]. Aspirin also interacts significantly with many other commonly used drugs and, whilst this can sometimes be used to clinical advantage, in other cases concomitant use is inadvisable or will require dosage adjustment or close monitoring.

Contraindications and precautions

Aspirin is absolutely contraindicated in patients with:

- haemorrhagic disorders, such as haemophilia
- recent history of gastrointestinal bleeding
- active gastric or duodenal mucosal lesions
- recent cerebral haemorrhage
- known aspirin hypersensitivity (or hypersensitivity to other non-steroidal anti-inflammatory agents).

In the setting of acute myocardial infarction or unstable angina, the definite and substantial benefits of aspirin therapy mean that it should not be withheld unless one of these contraindications is very clear [2]. After suspected acute ischaemic stroke, a CT scan is highly desirable to exclude haemorrhage and aspirin should preferably be withheld until the results are known. However, in circumstances where early CT scanning is not available, there is no evidence to support withholding aspirin if haemorrhagic stroke is thought unlikely on clinical grounds alone [3].

Relative contraindications, when aspirin should only be used very cautiously, include patients with dyspepsia and previous peptic ulceration and gastritis. Caution may also be justified in patients at increased risk of hypersensitivity reactions, such as those with severe late-onset persistent asthma [4] and those suffering from allergic rhinitis or chronic urticaria. It is best not to give aspirin to patients with gout, as low doses increase uric acid concentrations and may precipitate an acute attack, or to patients with known glucose-6-phosphate dehydrogenase deficiency, in whom it may lead to haemolysis.

High doses of aspirin are more likely than low doses to be associated with significant adverse effects, especially in elderly or dehydrated patients and those with renal or hepatic impairment. Regular monitoring of renal and hepatic function is advisable and measurement of plasma salicylate levels may also be helpful, ensuring that plasma salicylate concentrations do not rise above the threshold for toxicity of about 300 mg/ml (2.2 mmol/l). Cardiac failure may be exacerbated and become more difficult to control.

Use in particular patient groups

USE DURING PREGNANCY

Salicylates readily cross the placenta, but human studies have not demonstrated any significant risk or evidence of teratogenicity [5–7]. Furthermore, a meta-analysis of controlled trials of low-dose prophylactic aspirin in pregnancy did not show any significant excess of antepartum or postpartum haemorrhage or any other particular risks for mother or baby [8]. However, other than when used in low doses for prevention of pre-eclampsia or recurrent miscarriage, aspirin is best avoided during pregnancy, especially in the last trimester. High doses may prolong gestation and labour and be associated with increased maternal and neonatal bleeding, decreased birth weight, premature closure of ductus arteriosus and possibly persistent pulmonary hypertension in the newborn [9].

USE DURING BREAST FEEDING

As aspirin is excreted in breast milk, nursing mothers are advised not to take it. However, only chronic use of large doses of aspirin is likely to expose the child to significant amounts of salicylate [10].

USE IN CHILDREN

Since the recognition of the association between aspirin therapy and development of Reye's syndrome, aspirin has been contraindicated in children, especially those aged under 12 years, except for specific indications where the expected benefits outweigh the risks (e.g. treatment of Still's disease, Kawasaki disease).

USE IN OLDER PEOPLE

The absolute benefit of preventive treatment is likely to be greatest in patients most at risk. Many of the indications for low-dose aspirin therapy, such as ischaemic heart disease, stroke, colorectal cancer, cataract and cognitive decline, are age related and so most patients taking prophylactic aspirin will be elderly. Painful, inflammatory disorders are also common in old age. There is no significant age effect on the pharmacokinetics or response to aspirin, but older patients are more susceptible to gastrointestinal adverse effects and more prone to develop salicylism with high salicylate levels [11]. They are also more likely to have multiple pathology and polypharmacy which may increase their susceptibility to adverse effects. Closer monitoring of treatment may therefore be needed.

USE IN PATIENTS WITH RENAL FAILURE

A further deterioration in renal function may be associated with initiation of aspirin therapy in patients with pre-existing impairment. This is most likely in those who are salt or fluid depleted or who have renal problems associated with systemic lupus erythematosus, glomerulonephritis, cirrhosis or cardiac failure. These patients are reliant on renal prostaglandin synthesis to counteract the vasoconstrictive effects of the renin–angiotensin and adrenergic systems [12]. Caution is therefore appropriate and regular monitoring of renal function is required to detect any aspirin-induced deterioration. Low-dose aspirin may be indicated in dialysis patients to prevent thromboembolic complications.

USE IN PATIENTS UNDERGOING SURGICAL PROCEDURES

Aspirin treatment might be expected to be associated with an excess of haemorrhagic complications in surgical patients, but whilst some reports suggest increased perioperative bleeding and transfusion requirements [13], recent reports in patients taking low doses of aspirin are more reassuring [14,15]. Risk would seem to be significantly increased only when bleeding time is very prolonged (e.g. greater than eight minutes) and in these patients consideration should be given to stopping aspirin at least a week before surgery. Overall, however, there is little evidence that excessive bleeding is usually of significance for most operative procedures in most patients and the risk must be balanced against the potential benefits arising from protection against postoperative deep vein thrombosis and cardiovascular complications [16].

Despite the lack of good evidence to guide practice, many surgeons believe that aspirin increases perioperative bleeding to an unacceptable degree and recommend that aspirin is stopped a few days before elective surgical procedures. For example, in response to a questionnaire survey, most Dutch eye surgeons said that they discontinued aspirin therapy prior to cataract surgery, especially when using local anaesthesia, although they recognized that discontinuation was sometimes associated with serious systemic complications. The authors concluded that continued use of aspirin during surgery was to be recommended because of the possible life-threatening complications when it was stopped [17]. Most neuroanaesthetists surveyed about the continuation of low-dose aspirin in patients undergoing elective intracranial surgery felt that it was a risk factor for haemorrhagic complications, yet adopted no particular policy regarding its preoperative discontinuation [18].

Regional anaesthesia may possibly be associated with greater complication rates in patients on low-dose aspirin, with some studies reporting more frequent formation of extradural haematoma associated with epidural anaesthesia [19]. However, a large study in pregnant women receiving epidural blocks for pain relief during and after delivery did not find any increased risk of bleeding complications [20]. Patients taking low-dose aspirin will be more likely to develop a purpuric rash when tourniquets are used for nerve blocks and other procedures [21].

Particular surgical procedures may benefit from careful timing of aspirin dosing. For example, after coronary artery bypass surgery, antiplatelet therapy started preoperatively is associated with a small increase in drainage from the chest tube and in transfusion requirements and need for reoperation because of bleeding [22]. It is therefore generally recommended that treatment is started just after the vascular procedure. In microvascular surgery, failure is usually associated with thrombotic occlusion of a microvascular anastomosis or occasionally by flow impairment in the transferred or replanted tissue. In animal models, preoperative low-dose aspirin favourably influenced both processes [23], supporting its clinical use in free flap/reimplantation surgery.

Results from animal experiments suggest that administration of desmopressin acetate may be of value in reducing aspirin-associated perioperative bleeding when this is excessive, but clinical experience is limited [24].

Interactions with other drugs

Aspirin or its metabolites have a wide range of interactions with other drugs (Table 8.1), but relatively few are clinically important, especially when aspirin is being used in low dosage for its antithrombotic effect [25]. Many of the interactions are pharmacokinetic in nature [26]. There are no studies which have shown a significant interaction between any co-administered drug and the hydrolysis of aspirin, but there are numerous reports of drugs affecting its absorption and disposition. Pharmacodynamic interactions resulting in synergistic (addititive) or opposing (antagonistic) effects are less common.

No significant pharmacokinetic interaction has been shown between aspirin and paracetamol [27] or with opiates, such as codeine.

INTERACTIONS ARISING FROM ALTERED ABSORPTION

A number of agents affect the rate or extent of aspirin absorption, some increasing peak plasma drug concentration, such as antacids, caffeine, dipyridamole, metoclopramide and metoprolol, and others decreasing or delaying absorption, such as activated charcoal, cholestyramine, kaolin and griseofulvin.

Concurrent administration of aspirin with metoclopramide has been used in migraine sufferers to achieve pain relief [28]. Combination products of aspirin and caffeine are also available, though unlikely to offer any significant clinical advantages despite their widespread use. The interaction of antacids with enteric-coated preparations may in theory cause them to dissolve prematurely due to a rise in gastric pH, but positive confirmatory evidence of this seems to be lacking [25]. The concurrent administration of aspirin and dipyridamole results in higher plasma aspirin and salicylate concentrations and increased area under the curve, possibly by inhibition of aspirin esterase [29]. This may partly explain the apparent summative effects when very low-dose aspirin and dipyridamole are used together for thromboprophylaxis [30].

Activated charcoal has long been used in the treatment of salicylate poisoning, though as the binding is reversible [31], it is necessary to give repeated doses for maximum benefit. Early reports of enhanced aspirin absorption when taken concurrently with the histamine H2 receptor blockers cimetidine and ranitidine [32] have not been confirmed in subsequent studies, which instead suggest possible inhibition of salicylic acid metabolism [33].

Drug	Effect	Comment
Aspirin/salicylate affected		
Activated charcoal	Reduces absorption [85]	Potentially useful in treatment of salicylate toxicity
Alcohol	May inhibit aspirin esterase [64]. Both ulcerogenic	Increased gastric toxicity [67,68]
Antacids	Altered absorption and renal clearance [50,51,70,87]	Avoid within 1–2 h of each other
	Premature release of enteric-coated preparations [25]	May be clinically useful in salicylate toxicity
Caffeine	More rapid absorption [83]	Doubtful clinical significance
Cholestyramine	Delays absorption [84]	Avoid within 1–2 h of each other
Corticosteroids	Lower plasma salicylate [53]	Avoid unless essential
	Increased risk of GI toxicity [69]	
Dipyridamole	Increased absorption [29]. Additive antiplatelet effects	May be useful therapeutically [30]
Gold	Increased hepatotoxicity [79]	Avoid concurrent use
Griseofulvin	Reduces absorption [81]	Lower plasma salicylate levels
Metoclopramide	More rapid absoption [82]	May be useful therapeutically for earlier onset of analgesia in migraine [28]
Metoprolol	May increase absorption [86]	Doubtful clinical significance
Other drugs affected by aspirin/salicylate		
ACE inhibitors	Reduced effect, related to inhibition of prostaglandin synthesis [59–61]	Monitor clinical response
Acetazolamide	Protein binding displacement and decreased renal tubular secretion [37,38]	Avoid concurrent use
Alcohol	Inhibition of alcohol dehydrogenase [65]	Increased blood alcohol
Calcium channel blockers	Additive antiplatelet effects	Monitor clinical response
Methotrexate	Protein binding displacement and decreased renal tubular section [89]	Avoid concurrent use [52]
Midazolam	Protein binding competition	Shorter anaesthetic induction time [36]
Mifepristone	May affect efficacy	Avoid concurrent use (for at least 8–12 days)
Nitrates	Increased serum levels [62]	Avoid concurrent use
	Low-dose aspirin may reduce vasodilatory effects [63]	
	Additive antiplatelet effects [71]	

Drug	Effect	Comment
NSAIDs	Increased gastric toxicity and multiple pharmacokinetic interactions with individual agents [39–48]	Avoid concurrent use
Oral hypoglycaemics	Increased hypoglycaemic effect [77]	Monitor blood and adjust dose as needed
Phenytoin	Protein binding displacement [88]	Unlikely to be of clinical significance, but monitor free drug levels
Spironolactone	Reduced natriuretic effect [57]	Monitor clinical response
Thiopentone	Protein binding competition [36]	May prolong anaesthesia
Uricosuric agents	Reduced uricosuric effect	Avoid concurrent use or monitor clinical response
Valproic acid	Protein binding displacement and may inhibit metabolism [34,35]	Monitor free drug levels
Warfarin	Protein binding displacement. Increased risk of bleeding	Monitor INR – may sometimes be clinically useful [74,75]

Table 8.1
Drug interactions with aspirin and/or salicylic acid

INTERACTIONS ARISING FROM ALTERED DRUG DISPOSITION

The clearance of salicylic acid and its metabolites may be both induced or inhibited by co-administered drugs and, similarly, salicylate itself may influence the handling of some compounds.

Protein binding displacement of other highly bound drugs is the basis of many interactions. However, these are rarely of clinical significance as following displacement from plasma proteins, increased drug clearance occurs and although the total plasma concentration decreases, the concentration of unbound (active) drug is generally little altered. This is generally the case when aspirin is administered to patients receiving the anticonvulsant phenytoin, though more significant interaction leading to toxicity has been reported with valproic acid [34], possibly because of an additional inhibitory effect of salicylate on valproic acid metabolism [35]. Careful monitoring of free (unbound) plasma concentrations is therefore advisable if patients on anticonvulsants must take high-dose aspirin therapy. The anaesthetic effects of midazolam and thiopentone are increased in patients taking aspirin, probably because of competitive binding to plasma albumin [36]. This may result in shortened induction time and prolonged anaesthesia unless dosage adjustments are made. Clinically important displacement from plasma protein binding sites by the presence of high levels of salicylic acid has also been reported for the carbonic anhydrase inhibitor acetazolamide, with a greater than threefold increase in free drug levels [37]. Renal clearance of unbound acetazolamide is also decreased by salicylic acid, due to competitive inhibition of renal tubular secretion, and there are reports of serious toxicity arising from the drug combination [38], which should therefore be avoided.

Numerous pharmacokinetic interactions between aspirin and other non-steroidal anti-inflammatory drugs have been reported, most resulting in altered disposition of the co-administered drug rather than aspirin. A common mechanism is displacement from plasma protein binding sites by salicylate, leading to increased plasma clearance. This is considered to explain the interaction of aspirin with diclofenac [39], isoxicam [40], tolmetin [41], diflunisal [42] and the propionic acid derivatives flurbiprofen [43], ibuprofen [44], ketoprofen [45] and naproxen [46]. In contrast, protein binding of the propionic acid drug fenoprofen appears to be unaffected by salicylate, though total plasma clearance is increased, possibly due to induction of fenoprofen metabolism [47]. The interaction between indomethacin and aspirin appears to be more complex, with apparent reduction in indomethacin absorption, increased biliary clearance and enterohepatic recirculation and reduced renal clearance but no apparent effect on distribution or metabolic clearance of either drug. Overall, these effects compensate for each other and with repeated dosing there is generally little or no effect of aspirin on plasma concentrations of indomethacin [48]. Absence of reported interaction of aspirin with piroxicam [49] may reflect poor study design rather than a true exception to the rule.

On their own, none of these pharmacokinetic interactions between non-steroidal anti-inflammatory drugs and aspirin are likely to give rise to any clinically significant effects. Of more importance is the increased likelihood of the combination leading to serious gastric toxicity.

Drugs which alter urinary pH may affect the renal clearance of salicylic acid and the renal excretion of other acidic drugs may in turn be affected by salicylate. Antacids will result in a dose-dependent change in urinary pH and consequent increase in salicylate excretion. The effect of standard antacid doses is minimal, but high doses can increase renal clearance of salicylic acid as much as 12-fold, with highly significant reductions in plasma salicylate concentration which may be clinically valuable in treating salicylate poisoning [50,51]. Co-administration of aspirin and the highly acidic cytotoxic drug methotrexate significantly reduces the renal clearance of methotrexate, probably due to competitive inhibition of active tubular secretion. The plasma concentration of free methotrexate will be significantly raised and the combination has led to fatal pancytopenia [52]. The effects of uricosuric drugs such as probenecid and sulphinpyrazone may be diminished by aspirin, probably due to competition for renal routes of elimination.

Salicylic acid elimination is enhanced by corticosteroids, whether given orally [53] or by intra-articular injection [54]. Thus plasma salicylate concentrations in patients on high-dose, long-term aspirin treatment may be significantly lowered when steroid therapy is introduced and, conversely, salicylate toxicity may be precipitated when corticosteroids are discontinued. The underlying mechanism appears to be steroid-induced induction of salicylic acid glucuronidation and salicyluric acid formation [55]. Higher clearance of salicylic acid is also reported in women taking oral contraceptive steroids compared with women not using the pill [56].

The natriuretic and diuretic effects of spironolactone [57] and possibly loop diuretics such as frusemide and bumetanide [58] may be slightly reduced by high-dose aspirin, with potential worsening of concurrent heart failure. The activity of angiotensin-converting enzyme (ACE) inhibitors may also be reduced by aspirin, resulting in poor control of hypertension [59], but this is probably only of clinical importance when aspirin is being taken regularly. A single dose study failed to demonstrate any haemodynamic effects of aspirin in patients with chronic heart failure controlled with captopril [60], but there is some suggestion of interaction between ACE inhibitors and aspirin in the acute phase of myocardial infarction, the effect of ACE inhibition being less favourable among patients taking aspirin than among those not taking it at baseline [61]. While analgesic doses of aspirin may increase serum levels of glyceryl trinitrate, with greater chance of adverse side effects such as headache and hypotension [62], chronic low-dose aspirin therapy appears to reduce the vasodilatory effects of nitrates [63]. The importance of the interaction in clinical practice is uncertain.

Alcohol has been reported to inhibit aspirin esterase activity *in vitro* [64] and this may partly explain the greater gastric ulcerogenicity of aspirin when taken in combination with alcohol, there being a higher proportion of aspirin in the circulation than its less ulcerogenic metabolite, salicylate. No effect of alcohol on aspirin elimination *in vivo* has been demonstrated, but aspirin can increase blood alcohol levels through inhibition of alcohol dehydrogenase in the gastric mucosa [65].

Salicylic acid also competitively inhibits the metabolism of *m*-xylene, an industrial solvent which is widely used in the paint industry. In a volunteer study, co-administration of *m*-xylene and aspirin resulted in a significant reduction in renal excretion of glycine conjugates of both compounds [66]. This may be of clinical importance for patients on aspirin who are exposed to the solvent and may also lead to misinterpretation of the results of biological monitoring of workers who have recently taken the drug.

PHARMACODYNAMIC INTERACTIONS

Some of the effects of aspirin on the gastrointestinal tract are enhanced by alcohol, with greater likelihood of damage to the gastric mucosa (but not to the duodenum) and a small increase in gastrointestinal blood loss [67,68]. However, the increased risk is small and unlikely to be of importance in normal individuals taking moderate doses, but may become important with overuse. Concurrent use of aspirin and corticosteroids also increases the incidence of gastrointestinal bleeding and peptic ulceration [69]. Use of H2 receptor blockers and proton pump inhibitors may be protective.

Concurrent use of calcium channel blockers and aspirin has been reported to increase antiplatelet effects, occasionally resulting in significantly prolonged bleeding times and bruising [70]. The combination need only be avoided if the effects are clearly adverse. The antiaggregant effects of aspirin and glyceryl trinitrate [71] and aspirin and quinidine [72] also appear to be additive. The greater platelet antiaggregant effects of the combinations of aspirin with ticlopidine [73] or dipyridamole [30] may be used to therapeutic advantage.

Aspirin may enhance the effects of anticoagulant and thrombolytic drugs by a number of mechanisms, all leading to an increased tendency to haemorrhage. Their co-administration can offer therapeutic advantages in certain circumstances [74,75], but in general they should be taken together with great care. Displacement of warfarin from protein binding sites when aspirin treatment is initiated will produce a transient rise in free active drug and, conversely, stopping aspirin treatment may result in temporary under-anticoagulation. Large doses of aspirin may cause hypoprothrombinaemia, further enhancing the effects of warfarin [76].

The hypoglycaemic effects of the sulphonylurea hypoglycaemic agents may be increased by aspirin [77] and a reduction in insulin requirements has been described in patients with diabetes when high salicylate doses have been taken [78]. Neither effect is generally of any clinical importance with occasional or low-dose aspirin therapy. The precise mechanism involved is uncertain and it may be that the effects of the two drugs are merely summative.

In patients with rheumatoid arthritis taking regular high-dose aspirin, concurrent gold induction therapy increased aspirin-induced hepatotoxicity and use of another non-steroidal anti-inflammatory drug may be preferable [79].

There are anecdotal reports of occasional failure of intrauterine devices (IUDs) to prevent pregnancy due to concurrent use of aspirin [80]. It might be expected that aspirin would interfere with the action of mifepristone and it is recommended that its use should be avoided for at least 10–12 days.

KEYPOINTS

- Contraindications to aspirin are history of haemorrhagic disorders, such as haemophilia, gastrointestinal ulceration or bleeding, recent cerebral haemorrhage or known hypersensitivity.
- Especial care is appropriate in patients with dyspepsia and previous peptic ulceration and gastritis, those with late-onset persistent asthma, gout or renal or hepatic failure and during pregnancy.
- There is often concern about risk of excess haemorrhagic complications in surgical patients taking aspirin, but this is rarely of significance for most operative procedures and needs to be balanced against the potential benefits arising from protection against thrombotic complications.
- Aspirin or its metabolites have a wide range of interactions with other drugs, but relatively few are clinically important, especially when low doses are being taken. The antithrombotic effects of aspirin and anticoagulant and other antiplatelet drugs are often additive and can lead to serious bleeding complications.

REFERENCES

1. British National Formulary Number 34. London: British Medical Association and Royal Pharmaceutical Society of Great Britain, 1997: 192.
2. Collins R, Peto R, Baigent C, Sleight P. Aspirin, heparin, and fibrinolytic therapy in suspected acute myocardial infarction. N Engl J Med 1997; 336: 847–860.
3. Chen Z, Collins R, Peto R, Xie J-X, Liu L-S. Interpretation of IST and CAST stroke trials. Lancet 1997; 350: 445.
4. National Institute of Health Expert Panel Report. Guidelines for the Diagnosis and Management of Asthma. Bethesda: National Institutes of Health, 1997.
5. Slone D, Siskind V, Heinonen OP *et al.* Aspirin and congenital malformations. Lancet 1976; 1: 1373–1375.
6. Winship KA, Cahal DA, Weber JC, Griffin JP. Maternal drug histories and central nervous system anomalies. Arch Dis Child 1984; 59: 1052–1060.
7. Werler MM, Mitchell AA, Shapiro S. The relation of aspirin use during the first trimester of pregnancy to congenital cardiac defects. N Engl J Med 1989; 321: 1639–1642.

8. Leitich H, Egarter C, Husslein P, Kaider A, Schemper M. A meta-analysis of low-dose aspirin for the prevention of intrauterine growth retardation. Br J Obstet Gynaecol 1997; 104: 450–459.
9. De Swiet M, Fryers G. The use of aspirin in pregnancy. Br J Obstet Gynaecol 1990; 10: 467–482.
10. Findlay JW, DeAngelis RL, Kearney MF, Welch RM, Findlay JM. Analgesic drugs in breast milk and plasma. Clin Pharmacol Ther 1981; 29: 625–633.
11. Durnas C, Cusack GJ. Salicylate intoxication in the elderly: recognition and recommendations on how to prevent it. Drugs Aging 1992; 2: 20–34.
12. Day RO. Aspirin and salicylates. In: Kelley WN, Harris ED, Ruddy S (eds) Textbook of Rheumatology, 4th edn. New York: WB Saunders, 1994: 681–691.
13. Amrein PC, Ellman L, Harris WH. Aspirin-induced prolongation of bleeding time and perioperative blood loss. JAMA 1981; 245: 1825–1829.
14. Ala-Opas MY, Gronlund SS. Blood loss in long-term aspirin users during transurethral prostatectomy. Scand J Urol Nephrol 1997; 30: 203–206.
15. Billingsley EM, Maloney ME. Intraoperative and postoperative bleeding problems in patients taking warfarin, aspirin, and nonsteroidal antiinflammatory drugs. A prospective study. Dermatol Surg 1997; 23: 381–383.
16. Antiplatelet Trialists' Collaboration. Collaborative overview of randomised controlled trials of antiplatelet therapy – I: prevention of death, myocardial infarction, and stroke by prolonged antiplatelet therapy in various categories of patients. Br Med J 1994; 308: 81–106.
17. Koopmans SA, Van Rij G. Cataract surgery and anticoagulants. Documenta Ophthalmologica 1996; 92: 11–16.
18. James DN, Fernandes JR, Calder I, Smith M. Low-dose aspirin and intracranial surgery. A survey of the opinions of consultant neuroanaesthetists in the UK. Anaesthesia 1997; 52: 169–172.
19. Macdonald R. Aspirin and extradural blocks. Br J Anaesth 1991; 66: 1–3.
20. Sibai BM, Carilis SN, Thom E, Shaw K, McNellis D. Low-dose aspirin in nulliparous women: safety of continuous epidural block and correlation between bleeding time and maternal–neonatal bleeding complications. Am J Obstet Gynecol 1995; 172: 1553–1557.
21. Runcie CJ, Tansey P, Gordon G. Aspirin and intravenous regional blocks. Br J Hosp Med 1990; 43: 229–230.
22. Goldman S, Copeland JG, Moritz T *et al.* Starting aspirin therapy after operation: effects on early graft patency. Circulation 1991; 84: 520–526.
23. Peter FW, Franken RJ, Wang WZ *et al.* Effect of low dose aspirin on thrombus formation at arterial and venous microanastomoses and on the tissue microcirculation. Plastic Reconstruct Surg 1997; 99: 1112–1121.
24. Johnstone MT, Andrews T, Ware JA *et al.* Bleeding time prolongation with streptokinase and its reduction with 1–desamino-8–D-arginine vasopressin. Circulation 1990; 82: 2142–2151.
25. Stockley IH. Drug Interactions, 4th edn. London: Pharmaceutical Press, 1996.

26. Miners JO. Drug interactions involving aspirin (acetylsalicylic acid) and salicylic acid. Clin Pharmacokinet 1989; 17: 327–344.
27. Levy G, Regardh CG. Drug biotransformation interactions in man. V. Acetaminophen and salicylic acid. J Pharmaceut Sci 1971; 60: 608–611.
28. Ross-Lee LM, Heazlewood V, Tyrer JH, Eadie MJ. Aspirin treatment of migraine attacks: plasma drug level data. Cephalagia 1982;2: 9–14.
29. Nitelius E, Melander A, Wahlin-Boll E. Pharmacokinetic interaction of acetylsalicylic acid and dipyridamole. Br J Clin Pharmacol 1985; 19: 379–383.
30. European Stroke Prevention Study 2. Dipyridamole and acetylsalicylic acid in the secondary prevention of stroke. J Neurol Sci 1996; 143: 1–13.
31. Filippone GA, Fish SS, Lacouture PG, Scavone JM, Lovejoy FH. Reversible adsorption (desorption) of aspirin from activated charcoal. Arch Intern Med 1987; 147: 1390–1392.
32. Khoury W, Geraci K, Askari A, Johnson M. The effect of cimetidine on aspirin absorption. Gastroenterology 1979; 76; 1169.
33. Trnavska Z, Trnavsky K, Smondrk J. The effect of cimetidine on the pharmacokinetics of salicylic acid. Drugs Exp Clin Res 1985; 11: 703–707.
34. Goulden KJ, Doolley JM, Camfield PR, Fraser AD. Clinical valproate toxicity induced by acetylsalicylic acid. Neurology 1987; 37: 1392–1394.
35. Abbott FS, Kassam J, Orr JM, Farrell K. The effect of aspirin on valproic acid metabolism. Clin Pharmacol Ther 1986; 40: 94–100.
36. Dundee JW, Halliday NJ, McMurray TJ. Aspirin and probenecid pretreatment influences the potency of thiopentone and the onset of action of midazolam. Eur J Anaesthesiol 1986; 3: 247–251.
37. Sweeney KR, Chapron DJ, Brandt JL *et al.* Toxic interaction between acetazolamide and salicylate: case reports and a pharmacokinetic explanation. Clin Pharmacol Ther 1986; 40: 518–524.
38. Cowan RA, Hartnell GG, Lowdell CP, Baird IM, Leak AM. Metabolic acidosis induced by carbonic anhydrase inhibitors and salicylates in patients with normal renal function. Br Med J 1984; 289: 347–348.
39. Willis JV, Kendall MJ, Jack DB. A study of the effect of aspirin on the pharmacokinetics of oral and intravenous diclofenac sodium. Eur J Clin Pharmacol 1980; 18: 415–418.
40. Esquival M, Cussenot F, Olgilvie RI, East DS, Shaw DH. Interaction of isoxicam with acetylsalicylic acid. Br J Clin Pharmacol 1984; 18: 567–571.
41. Cressman WA, Wortham GF, Plostnicks J. Absorption and excretion of tolmetin in man. Clin Pharmacol Ther 1976; 19: 224–233.
42. Schultz P, Perrier CV, Ferber-Perret F, Vanden Heuvel WJA, Steelman SL. Diflunisal, a new-acting analgesic and prostaglandin inhibitor: effect of concomitant acetylsalicylic acid therapy on ototoxicity and on disposition of both drugs. J Int Med Res 1979; 7: 61–68.
43. Brooks PM, Khong TK. Flurbiprofen–aspirin interaction: a double-blind crossover study. Curr Med Res Opin 1977; 5: 53–57.

44. Grennan DM, Ferry DG, Ashworth ME, Kenny RE, Mackinnon M. The aspirin–ibuprofen interaction in rheumatoid arthritis. Br J Clin Pharmacol 1979; 8: 497–503.
45. Williams RL, Upton RA, Bushkin JN, Jones RM. Ketoprofen–aspirin interactions. Clin Pharmacol Ther 1981; 30: 226–231.
46. Segre EJ, Chaplin M, Forchielli E, Runkel R, Sevelius H. Naproxen–aspirin interactions in man. Clin Pharmacol Ther 1974; 15: 374–379.
47. Rubin A, Rodda BE, Warrick P, Gruber CM, Ridolfo AS. Interactions of aspirin with non-steroidal anti-inflammatory drugs in man. Arthritis Rheumatol 1973; 16: 635–645.
48. Kwan KC, Breault GO, Davis RL *et al.* Effects of concomitant aspirin administration on the pharmacokinetics of indomethacin in man. J Pharmacokinet Biopharm 1978; 6: 451–476.
49. Hobbs DC, Twomey TM. Piroxicam pharmacokinetics in man: aspirin and antacid interaction studies. J Clin Pharmacol 1979; 19: 270–281.
50. Levy G, Leonards JR. Urine pH and salicylate therapy. JAMA 1971; 217: 81.
51. Hansten PD, Hayton WL. Effect of antacid and ascorbic acid on serum salicylate concentration. J Clin Pharmacol 1980; 30: 326–331.
52. Mandel MA. The synergistic effect of salicylates on methotrexate toxicity. Plast Reconstruct Surg 1976; 57: 733–737.
53. Klinenberg JR, Miller F. Effect of corticosteroids on blood salicylate concentration. JAMA 1965; 194: 131–134.
54. Edelman J, Potter JM, Hackett LP. The effect of intra-articular steroids on plasma salicylate concentrations. Br J Clin Pharmacol 1986; 21: 301–307.
55. Graham GG, Champion GD, Day RO, Paull PD. Patterns of plasma concentrations and urinary excretion of salicylate in rheumatoid arthritis. Clin Pharmacol Ther 1977; 22: 410–420.
56. Miners JO, Grgurinovich N, Whitehead AG, Robson RA, Birkett DJ. Influence of gender and oral contraceptive steroids on the metabolism of salicylic acid and acetylsalicylic acid. Br J Clin Pharmacol 1986; 22: 135–142.
57. Tweeddale MG, Ogilvie RI. Antagonism of spironolactone-induced natriuresis by aspirin in man. N Engl J Med 1973; 289: 198.
58. Kaufman J, Hamburger R, Mathheson J, Flamebaum W. Bumetanide-induced diuresis and natriuresis: effect of prostaglandin synthetase inhibition. J Clin Pharmacol 1981; 21: 663–667.
59. Moore TJ, Crantz FR, Hollenberg NK *et al.* Contribution of prostaglandins to the antihypertensive action of captopril in essential hypertension. Hypertension 1981; 3: 168–173.
60. Van Wijngaarden, Smit AJ, De Graeff PA *et al.* Effects of acetylsalicylic acid on peripheral haemodynamics in patients with chronic heart failure treated with angiotensin-converting enzyme inhibitors. J Cardiovasc Pharmacol 1994; 23: 240–245.
61. Nguyen KN, Aursnes I, Kjekshus J. Interaction between enalapril and aspirin on mortality after acute myocardial infarction: subgroup analysis of the

Cooperative New Scandinavian Enalapril Survival Study II (CONSENSUS II). Am J Cardiol 1997; 79: 115–119.

62. Weber S, Rey E, Pipeau C *et al.* Influence of aspirin on the haemodynamic effects of sublingual nitroglycerin. J Cardiovasc Pharmacol 1983; 5: 874–877.
63. Key BJ, Keen M, Wilkes MP. Reduced responsiveness to nitro-vasodilator following prolonged low dose aspirin administration in man. Br J Clin Pharmacol 1992; 34: 453–454P.
64. Builder J, Landecker K, Whitecross D, Piper DW. Aspirin esterase activity of gastric mucosal origin. Gastroenterology 1977; 73: 15–18.
65. Roine R, Gentry T, Hermandez-Munoz R, Barona E, Lieber CS. Aspirin increases blood alcohol concentrations in humans after ingestion of alcohol. J Am Med Assoc 1990; 264: 2406–2408.
66. Campbell L, Wilson HK, Samuel AM, Gompertz D. Interactions of m-xylene and aspirin metabolism in man. Br J Indust Med 1988; 45: 127–132.
67. Goulston K, Cooke AR. Alcohol, aspirin and gastrointestinal bleeding. Br Med J 1968; 4: 644.
68. Lanza FL, Royer GL, Nelson RS, Rack MF, Seckmann CC. Ethanol, aspirin, ibuprofen and the gastroduodenal mucosa: an endoscopic assessment. Gastroenterology 1985; 80: 767–769.
69. Carson JL, Strom BL, Schinnar R *et al.* Do corticosteroids really cause upper GI bleeding? Clin Res 1987; 35: 340A.
70. Verzino E, Kaplan B, Ashley JV, Burdette M. Verapamil–aspirin interaction. Ann Pharmacother 1994; 28: 536–537.
71. Karlberg KE, Ahlner J, Henriksson P, Torfgard K, Sylven C. Effects of nitroglycerin on platelet aggregation beyond the effects of acetylsalicylic acid in healthy subjects. Am J Cardiol 1993; 71: 361–364.
72. Lawson D, Mehta J, Mehta P, Lipman BC, Imperi GA. Cumulative effects of quinidine and aspirin on bleeding time and platelet alpha-2-adrenoreceptors: potential mechanism of bleeding diathesis in patients receiving the combination. J Lab Clin Med 1986; 108: 581 586.
73. Strano A, Davia G, Cannizzaro S, Timpone S, Alaimo P. Effects of administration of ticlopidine and low dose aspirin on platelet function in IHD patients. Thromb Haemostas 1985; 54: 394.
74. ISIS-2 (Second International Study of Infarct Survival) Collaborative Group. Randomised trial of intravenous streptokinase, oral aspirin, both, or neither among 17,187 cases of suspected acute myocardial infarction. ISIS-2. Lancet 1988; 2: 349–360.
75. Cappelleri J, Fiore L, Brophy M, Deykin D, Lau J. Efficacy and safety of combined anticoagulant and antiplatelet therapy versus anticoagulant monotherapy after mechanical heart valve replacement: a meta-analysis. Am Heart J 1995; 130: 547–552.
76. Park PK, Leck JB. On the mechanisms of salicylate-induced hypothrombinaemia. J Pharm Pharmacol 1981; 33: 25.

77. Richardson T, Foster J, Mawer GE. Enhancement by sodium salicylate of the blood glucose lowering effect of chlorpropamide–drug interaction or summation of similar effects? Br J Clin Pharmacol 1986; 22: 43–48.
78. Reid J, Lightbody TD. The insulin equivalence of salicylate. Br Med J 1959; 1: 897.
79. Davis JD, Turner RA, Collins RL, Ruchte IR, Kaufmann JS. Fenoprofen, aspirin and gold induction in rheumatoid arthritis. Clin Pharmacol Ther 1977; 21: 52–61.
80. Buhler M, Papiernik E. Successive pregnancies in women fitted with intrauterine devices who take anti-inflammatory drugs. Lancet 1983; i: 483.
81. Phillips KR, Wideman SD, Cochran EB, Becker JA. Griseofulvin significantly decreases serum salicylate concentrations. Paediatr Infect Dis J 1993; 12: 350–352.
82. Volans GN. The effect of metoclopramide on the absorption of effervescent aspirin in migraine. Br J Clin Pharmacol 1975; 2: 57–63.
83. Yoovathaworn KC, Sriwatanakul K, Thithapandha A. Influence of caffeine on aspirin pharmacokinetics. Eur J Drug Metab Pharmacokinet 1986; 11: 71–76.
84. Hahn KJ, Eden W, Schettle M *et al.* Effect of cholestyramine on the gastrointestinal absorption of phenprocoumon and acetylsalicylic acid in man. Eur J Clin Pharmacol 1972; 4: 142–145.
85. Levy G, Tsuchiya T. Effect of activated charcoal on aspirin absorption in man. Part 1. Clin Pharmacol Ther 1972; 13: 317–322.
86. Spahn H, Langguth P, Kirch W, Mutschler E, Ohnhaus F.E. Pharmacokinetics of salicylates administered with metoprolol. Arzneim Forsch 1986; 36: 1697–1699.
87. Hansten PD, Hayton WL. Effect of antacid and ascorbic acid on serum salicylate concentration. J Clin Pharmacol 1980; 30: 326–331.
88. Leonard RF, Knott PJ, Rankin GO, Robinson DS, Melnick DE. Phenytoin–salicylate interaction. Clin Pharmacol Ther 1981; 29: 56–60.
89. Liegler DG, Henderson ES, Hahn MA, Oliverio VT. The effect of organic acids on renal clearance of methotrexate in man. Clin Pharmacol Ther 1969; 10: 849–857.

Index

Page numbers appearing in *italic* refer to tables, page numbers appearing in **bold** refer to figures